DAMAGE IDENTIFICATION OF NONLINEAR BREATHING CRACKS UNDER DIFFERENT EXCITATIONS

by

Hajira Aftab

ABSTRACT

Structures in real life scenarios like industrial robots, spaceships, aircrafts, submarines, bridges and pressure vessels undergo different loading conditions. These structures are prone to damage due to the risks related to fatigue and ageing. Small cracks may appear in these complex structures and prove to be fatal. These cracks open and close under load and thus are known as breathing cracks. Breathing cracks exhibit nonlinear behavior. Various nonlinear techniques exist for detection and localization of breathing crack in one-dimensional structures such as beams. However, in plate-like (2D) structures these types of cracks are difficult to identify. Researchers have modeled open part-through cracks analytically in plates, but it still needs to be explored in case of breathing part-through cracks. In this thesis, a mathematical model of a breathing part-through crack in a plate based on piecewise equations has been presented. The cracked plate has been modeled as a bilinear oscillator with two frequencies corresponding to two stiffness regions, indicating breathing. The piecewise equations used to model breathing crack has been combined by considering its dynamic characteristics at the stiffness interface. The method of multiple scales has been employed to find out the solution of the cracked plate, using different excitations and enhanced to find out breathing crack frequency. The breathing frequency obtained has been validated through mathematical equations, numerical simulations as well as by experimental work. Further, in order to identify breathing crack, the Hilbert transform has been applied on 2D structure and the crack detected by utilizing the response of the plate. The appearance of nonlinearities in the response spectrum helps in crack detection which depends primarily on the type of excitation. It has been observed that Impulse excitation give good identification and the Harmonic excitation gives good severity estimation. Random excitations have also been exploited in this research to assess if better results are achievable regarding identification and severity of crack. The results indicate that the proposed method successfully identifies the nonlinear breathing crack via all excitations. Another important finding is that, the instantaneous frequency obtained can be related to severity estimation of the breathing crack. The presented research uses the attributes of the various excitation signals and enhances the detection and severity estimation techniques. Furthermore, the presented method does not require the base-line data for identification of breathing crack so it can be called as a base-line free method.

Keywords: Breathing part-through crack, Damage identification, Excitations, Plate

TABLE OF CONTENTS

LIST OF TABLES

NOMENCLATURE

List of Symbols

α Real phase at first order of perturbation theory

α_{bb} Non dimensional bending compliance coefficients

α_{bt} Non dimensional stretching-bending compliance coefficients

α_{tt} Non dimensional stretching compliance coefficients

$\Delta\omega_\%$ Percentage Frequency Change

$\epsilon_{1,2}$ Nonlinear cubic spring stiffness; 1=open crack; 2=close crack

Γ Complex amplitude at zeroth order of perturbation theory

λ Real amplitude at first order of perturbation theory

μ Damping co-efficient

Ω Excitation frequency

ω_c Natural frequency of the plate with closed crack/intact

ω_o Natural frequency of the plate with open crack

ω_r Random input frequency

ω_{br} Natural frequency of the plate with breathing crack

ρ Density of plate

ε^n Perturbation parameter for the method of multiple scales,n=0: fast time scales; n=1:slow time scale

$\varepsilon_x,\varepsilon_y$ Middle surface strain

A_{mn} Arbitrary amplitude

G_{mn} Complex Modal Component-3

k_c Stiffness of the plate with close crack/intact

k_o Stiffness of the plate with open crack

K_{mn} Complex Modal Component-1

M_{mn} Complex Modal Component-2

n_x, n_y, n_{rs} In-plane forces per unit length

P_z Point Load

P_{mn} Complex Modal Component-4

U_{mn} Time dependent modal coordinate

X_m, Y_n Modal functions of the cracked plate

v Poisson's ratio

a Half crack length

b Breadth of the plate

D Flexural Rigidity

E Modulus of Elasticity

h Thickness of plate

l Length of the plate

w Transverse deflection

List of Abbreviations

BCs Boundary Conditions

$CCFF$ Clamped-Clamped-Free-Free

$CCSS$ Clamped-Clamped-Simply supported-Simply supported

$CFFF$ Clamped-Free-Free-Free

CPT Classical Plate Theory

CWT Continuous Wavelet Transform

FFT Fast Fourier Transform

HT Hilbert Transform

IF Instantaneous Frequency

IFR Instantaneous Frequency Ratio

LSM   Linear Spring Model

SHM Structural Health Monitoring

SSSS All Sides Simply Supported

1.INTRODUCTION

1.1   INTRODUCTION

Structural Health Monitoring (SHM) has become one of the most widely researched area due to its significance in fields like mechanical, mechatronics, robotics, civil, aerospace, industrial, and nuclear engineering. Many engineering structures that have been developed in the past are still in use regardless of their over-aging and exceeding design life, due to economic reasons. Structures in real life scenarios undergo different loading conditions which cause small cracks to appear in them. Although small, these cracks pose danger as they may lead to catastrophic failure to the whole structure, if they are not detected. Hence, extensive study has been carried out on the subject. When load is applied on such structures these small cracks open and close, exhibiting breathing phenomenon, thus inducing nonlinearity in the response output. In addition to these macro scale engineering structures, breathing cracks in micro structures such as Micro-Electro-Mechanical Structures (MEMS), are also very common and an issue for which reliability tests are carried out. These cracks in MEMS devices affect the resonant frequency and electrical resistance, degrade the sensor output which might lead to their failure [1]. Thus, it is a need of the hour that good and robust techniques should be developed for improved damage detection. Recent advancements in data acquisition devices, signal processing methods and computer technology has resulted in cost-effective research techniques in SHM for engineering structures [2].

1.2 MOTIVATION

The real-life scenarios are depicted more accurately by nonlinear models instead of linear models. Therefore, the trend of SHM has shifted from linear to nonlinear, in the hope of getting more effective results for damage detection. Breathing cracks appear in structures at an early stage and can only be detected by nonlinear models. The proposed research deals with the development of nonlinear methodology using advanced signal processing for detecting breathing cracks using varying excitations. This research is expected to help prevent failure of structure by advance warning, hence providing efficient monitoring and maintenance.

Mainly, two plate theories have been developed in the literature known as the Kirchoff plate theory (for thin plates) and the Mindlin Reissner plate theory (for moderate thick plate). These theories have been used by researchers to formulate cracked plate equations by adding a Line Spring Model (LSM) for the crack. Mathematical methods for plates having center part-through crack, crack at different location and orientation, all-over crack, through crack and internal crack have been developed. These theories are applicable to open cracks whereas the dynamical behavior of a breathing crack has been addressed mathematically by only a few researchers and that too, only in the case of an all-over part-through crack and through crack [3, 4]. The role of different excitation signals in the response of the system also needs to be investigated.

Therefore, the motivation for the research work was to extend the existing equation for open part-through crack to breathing part-through crack for different excitations. The method utilizes the nonlinear response from breathing crack plate for various inputs. Furthermore, the study investigates different boundary conditions(BCs) for the plate, based on Leissa classical plate theory to check the effectiveness of the method.

1.3 RESEARCH OBJECTIVES

It is the need of the hour that advanced signal processing techniques should be applied to the response of the damaged model to obtain the desired results. For this, a deep understanding of the cracked plate model is required. As the crack is breathing, so a bilinear oscillator model have been considered which gives two stiffness values for two states (open state and closed state). The vibration analysis of open part-through crack by another researcher gives a detailed derivation of the crack plate equation [5]. However, the concept has been extended to breathing part-through crack to find out the breathing crack frequency in this thesis by using the impulse, harmonic as well as random excitation. The research in the thesis seeks to:

1. Develop a mathematical model for the identification of a nonlinear part-through breathing crack in a rectangular plate. The study investigates the response equations using impulse, harmonic and random excitation.

2. Make the method base-line free by assuming frequency at closed crack state equal to intact plate frequency.

3. Estimate the severity of the crack.

4. Apply the Hilbert Transform (HT) to the response of the breathing crack model.

5. Define terms like minimum difference percentage ($\Delta\omega\%$) and Instantaneous Frequency Ratio (IFR) for comparison of results.

6. Investigate the trends for different BCs of a plate.

7. Implement the method on the response from simulated experimental data from ANSYS to check results for various BCs and excitations.

8. Verify the proposed theory with actual experimentation.

1.4   FLOWCHART OF THE RESEARCH PROCEDURE

The flowchart in Figure 1.1 shows the path adopted for carrying out the research.

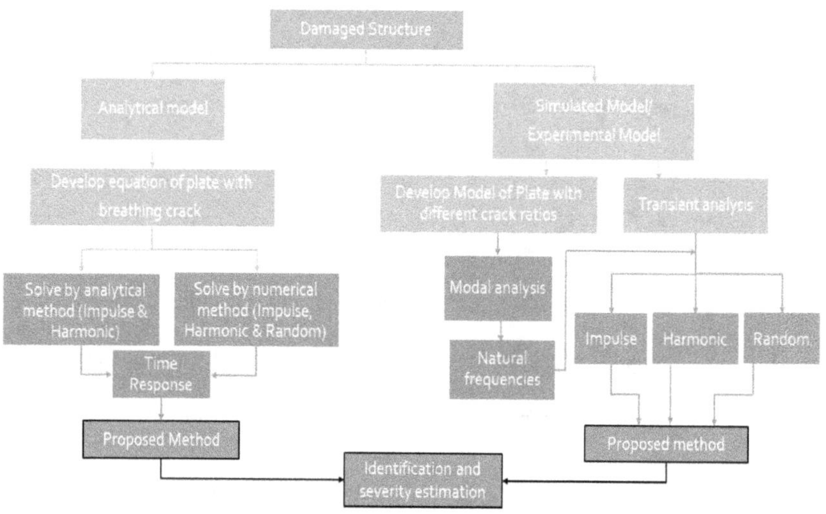

Figure 1.1: Flowchart of research procedure

The research procedure consists of two main streams. The analytical model and the simulated model. The analytical modelling was done for part-through crack with four different BCs. The model was solved for impulse excitation and harmonic whereas for random input numerical model was adopted. The proposed methodology was applied

4

to the time response for identification and severity estimation.

Similarly, for simulated model four BCs were considered. The modal analysis was carried out in order to obtain the natural frequencies and these were used in transient analysis to find the range of first five modes. The modal analysis was performed in parallel to the analytical modeling. The transient analysis was performed by three excitations and the proposed methodology was applied to the response to identify the crack. Lastly, experimental investigation was carried out on a cantilever plate to validate the study.

1.5 RESEARCH CONTRIBUTIONS

The research contributions related to this thesis are as follows:

1. Identification and severity estimation of a breathing crack via nonlinear dynamics.[6]

2. Experimental investigation of a breathing crack under different excitations.

1.6 THESIS ORGANIZATION

Chapter 2 presents the literature review of the plate theories, different crack models in plates and breathing cracks theories in case of beams, specifically in reference to vibrational study. It also points out research regarding different excitations for the cracked structures. The signal processing techniques used by various researchers for crack detection has also been discussed.

Chapter 3 provides the detailed mathematical formulation of the breathing crack model from the existing equations. It also shows frequencies of open and breathing part-through cracks obtained analytically for various excitations i.e. impulse, harmonic and random.

Chapter 4 discusses the solution methodologies presented in this research that can be applied to the derived nonlinear piecewise equations. It also shows the application of the presented HT on the response signal. Tables and figures show the significance of the study.

Chapter 5 gives a detailed account of the simulated finite element model of the breathing crack in a plate using software ANSYS. It also lists the breathing frequencies obtained by simulations of the plate under different BCs and excitations. Hence, validating the results obtained analytically.

Chapter 6 provides the experimental setup and results of the plate having breathing crack with different severities. Thus, authenticating the study. Furthermore, comparison of analytical, simulated and experimental work is presented and the results are established.

Chapter 7 concludes the thesis with useful analysis for breathing crack in a plate. It gives suggestions regarding potential future work.

2.LITERATURE REVIEW

This chapter covers the historical background on the damage identification in plate-like structures which comes under the area of SHM and the solution methodologies associated with the nonlinear equations. It also provides an introduction to the signal processing techniques that have been employed for damage detection by means of different excitations.

2.1 HISTORICAL BACKGROUND

Over the last two decades SHM is under constant progression in research and engineering areas i.e. from only monitoring to self-healing ability [2]. In 1993, Rytter defined levels of SHM which are the basis for this research [7] i.e. Level 1: Presence of damage; Level 2: Location of damage; Level 3: Severity of damage and Level 4: Prediction of remaining useful life of the damaged structure. Many methods like vibrations, ultrasonic testing, thermal testing, liquid penetrant testing, electromagnetic testing etc. have been developed by different researchers for the different types of damage detection [8]. However, the most popular are the vibrations and ultrasonic methods for engineering structures. There are two ways to proceed in literature for damage detection using the above mentioned methods i.e. linear and nonlinear. The linear methods although give good indications of macro cracks but literature reveals that the presence of micro cracks, which are more likely to occur due to fatigue or cyclic loading in real structures, are better identified by nonlinear methods [9, 10]. In ultrasonic guided waves, linear methods use parameters like attenuation, transmission and reflection [11] and the nonlinear methods use nonlinearities in harmonics produced by the damage itself [12]. The nonlinearities introduced by the damage depends on the type of damage that is to be considered or the environmental conditions of the structure. According to a review, the types of nonlinearities include breathing effect, clapping, hysteresis [13] rough surfaces contact and other dissipative nonlinearities [14]. The breathing crack is a classical nonlinear model, based on the opening and closing of crack under loading [15, 16, 17, 18]. The common methods employed for breathing cracks are the higher harmonic generation [19], frequency mixing [20, 21, 22, 23] and time-frequency [24, 25] as mentioned in literature. Recently, the hybrid techniques of vibro-acoustics are being used for the detection of damage arising due to breathing cracks [26, 27]. It not only incorporates wave propagational study but also vibrational study of beams.

Vibration is the earliest known method to detect damage in structures. When the structure undergoes some change, its natural frequency and mode shapes changes due to change in physical properties (stiffness, mass and damping) [28, 29]. Studies reveal that change in natural frequency is a global indicator that can be obtained from a single measurement point of a structure [30, 31, 32].

A breathing crack can be modeled based on bilinear stiffness i.e. the change in stiffness in two states (open and close), which causes the natural frequency to change too. Initially, researchers considered only open cracks for detection of damage but later it was found that the natural frequencies for open and breathing cracks are different. Hence, open cracks cannot be assumed in case of breathing cracks [33, 34, 35]. The natural frequencies decrease as function of crack length due to opening and closing of crack, but the drop is very low as compared to open cracks. So, it is not a good indicator of breathing crack. However, harmonics produced alongside the natural frequency are acceptable nonlinear damage indicators [16, 34, 36]. This eliminates the need to use baseline data for detection of damage but for quantification of crack, some other technique must be applied like the HT, statistical entropy, wavelet transform etc.

2.2 PLATE STRUCTURES AND THEORIES

Plate, shells and beams are the basic elements for structural analysis. The thickness of a plate is very less as compared to its width and length. The body of aircrafts, submarines, reactors, pressure vessels, printed circuit boards, solar panels, rotors, turbine blades etc. can be modeled by considering a plate-like structure. There are four main categories for theory of plates depending on the thickness of the plate i.e. Kirchhoff's plate (for thin plates), Mindlin-Riessner's plate (for moderate thickness plates), Reddy's plate (for thick plates) and three dimensional elasticity theory (for 3-D plates). The Kirchhoff's plate is also called the classical plate theory and is used for thin plates and shells. The Mindlin-Riessner's plate theory, based on the Timoshenko elastic beam theory, connects to the Lord Rayleigh approximate theory on plates and is the first to establish the relationship between the two to find out the shear factor used in Mindlin theory. Vibrational study of plates have been an interesting topic for researchers for the last two centuries. However, its application on structures and noise was introduced towards the end of 19th century [37]. Initially, the membrane theory of plates was formulated and later on, the governing equation for free vibrational analysis of plate was formed [38]. Another differential equation using an energy approach was formulated in 1887 [39]. Researchers have used different loading and BCs to find out the deflection in plates [40]. In a comprehensive work on plate theory, the approximate natural frequency formulas for 21 types of BCs by applying different combinations, along with nodal patterns, size and material properties was presented [41]. It was shown in another research, that the BCs and geometries of plates greatly effects the plate analysis methods as closed form solutions are obtained only for a specific set of geometries [42]. The flexural vibrations of an isotropic Mindlin plate was investigated by another researcher and the numerical solution for a simply supported square plate obtained. In 1973, the exceptional work done by Liessa on Classical Plate Theory (CPT) covered almost everything on the vibration of thin plates and gave exact results [43]. Other studies used the Rayliegh-Ritz method to predict the natural frequencies of flexural vibration of plates based on Mindlin plate theory [44]. The vibrational analysis of thick plates with different aspect ratios of size, thickness and boundary conditions was carried out in another research by employing the Rayliegh-Ritz method [45]. Finite element methods have also been used to find deflection in plates [37].

The research work in this thesis is based on Kirchhoff's plate theory (CPT) as an isotropic, thin plate has been considered. It will be a step forward in the development of an improved model for structures that are light, reliable and efficient.

2.3 PLATE MODELS WITH DAMAGE

Many studies on plates with different types of damages have been researched in the past few decades i.e. through crack, part-through crack, all-over crack, central crack and oblique crack. However, two main categories are modeled according to the crack dynamics i.e. (i) open crack model and (ii) breathing crack model.

2.3.1 Open Crack Model

In the early seventies, work on cracked plates was initiated and an equation for the case of simultaneous stretching and bending on a discontinuity present at the center of the plate was formulated [46]. The part-through crack was modeled as a Linear Spring Model (LSM) using Kirchhoff's plate theory. Later on, the work was extended for arbitrarily location of crack on plate [47]. The work was further enhanced by considering the Stress Intensity Factors (SIF) at the crack edges giving rise to Modified Linear Spring Model (MLSM) in which the opening angle of the Riessner plate was equated to the rotation of the CPT. This research proved that SIF of Reissner-plate and Kirchoff-plate was same for very shallow cracks [48]. In another research, a new LSM was developed by coupling boundary element method to find surface flaws in plate [49]. The method facilitated stress to be calculated at the crack and at the far end of the cracked plate. During the last twenty years, the work on cracked plate have been modelled by many researchers considering different parameters. Crack has been investigated for different BCs, different loading conditions and different orientations [4, 5, 50].

2.3.2 Breathing Crack Model

The breathing crack can be modeled on the basis of change in stiffness during its opening and closing state forming a bi-linear stiffness model [51]. If the rubbing or rough surface contact is considered, the nonlinearities are produced by these phenomena [52]. It was theorized that this model was a nonlinear oscillator due to generation of subharmonics produced by the existence of a certain level of displacement amplitude due to rubbing [53].

Another model known as the nonlinear relaxator has also been developed which is known as bi-stable model. This model considered the effective force which varied when the breathing crack opens and closes. It is useful for describing the hysteretic nonlinearity [54].

When the structure having a breathing crack is under tension, the crack opens while under compression the crack closes. The closed state will allow the excitation wave to pass as if the structure is intact while the open state will behave as an open crack. The crack will have a bilinear frequency and a bilinear stiffness.

The breathing crack model has also been represented as function of smoothly varying stiffness and represented by a polynomial of higher order using the Volterra series [55]. Change in flexibility has also been used to model the breathing crack [56]. Most studies reveal that many analytical models for various breathing cracks in a beam have been developed but the models for breathing crack in a plate are very few. An all-over breathing crack has been modeled by a researcher for thin plate and moderate thick plate [4, 3]. The thesis focuses in developing an equation for a centrally located part-through breathing crack.

2.4 BREATHING CRACK IDENTIFICATION METHODS

The breathing crack identification depends on a number of factors. It depends heavily on the type of excitation. Also, the analysis of response signal for nonlinearities play a major role in the breathing crack characterization. So, one must know the nonlinear signal processing techniques.

2.4.1 Impulse Excitation

In literature, impulse excitation has been used for severity estimations and baseline-free detection methods. Several researchers have used different techniques with impulse excitation like time-frequency method [24, 57], drop in natural frequency [58] , comparison of natural frequencies at a single point from damaged and intact beam [19] and Continuous Wavelet Transform (CWT) [59] to detect crack. Some have even worked on evaluating the crack by using tonal bursts and observing the increase in normalized frequency and amplitude with crack severity ratio [60]. Whereas, some studies have used nonlinearity as a measure of crack severity for offset boundary conditions [61]. Multi-crack detection based on natural frequencies using Hilbert Huang Transform (HHT) in non-uniform cross sectional beam has also been explored [62]. Lamb waves excitations have been used and a time reversal operator (TRO) that was only decomposed at the second-harmonic frequency has been studied [63]. The number of significant eigenvalues were found to be exactly equal to the number of fatigue cracks. However, in most of the studies baseline data has been used. For plate-like structures, a Spectral Damage Index (SDI) based on nonlinearities [12] and spectral correlation in the presence of noise [64] arising from Lamb waves have been proposed. The method successfully detects and quantifies the crack on aluminum plate, but the healthy plate data was required.

2.4.2 Harmonic Excitation

Excitation signals play an important role in the damage detection scheme. The response signals from various excitation signals have been explored using different techniques. In the early 2000's, the behavior of breathing crack in a beam was studied showing that breathing crack exhibits nonlinear response depending on the type of excitation [65, 66]. The cracks on a steel beam were detected by investigating the generated super-harmonics [67]. Later on, an analytical method that estimated crack location and severity on an Aluminum beam, concerning to damping effect was proposed [68]. However, the super-harmonics produced in the response spectrum posed a limitation, as these could also be generated by electrical equipment or sensors. This drawback was addressed in another research, by investigating sub-harmonics that were also found in the response spectrum at some threshold frequency [69]. This method eliminated the interference from electrical equipment but lacked severity estimation. Combination of two waves at different frequencies was then used to detect and locate crack [70]. Another study [71], based on the combinational tone analysis, used a third frequency which originated in the response signal alongside the two harmonic signals, to locate crack. It removed the discrepancies of using super- and sub-harmonics but required further improvement in detecting smaller cracks and estimating their severity. The level of excitation of harmonics have also been discussed by various researchers to estimate severity [55]. Different methods like modal curvature shapes [16] and the effect on super- and sub-harmonics produced in the response spectrum, due to gravitational force and excitation force amplitude, at various angles has also been explored in case of a slant crack [72]. It was found that the gravitational effects were more prominent under low level of loads whereas higher loads diminished the gravitational force effect. In addition to transverse and slant cracks, finite element modeling of three different types of breathing cracks have also been reported [61] i.e. non-penetrating parabolic crack (NPPC), penetrating trapezoid crack (PTC), and uniform-penetrating crack (UPC). The combined effect of different spectrum cascades like acceleration, velocity and displacement phase portraits were observed and damage location and severity was found. The detection of cracks has not been limited to beams only but plates have also been considered. The plate-like structures have been excited by using low-profile piezoceramic sensors and successfully detected breathing crack [73, 74, 75]. In addition to crack detection, the crack severity has also been found by using amplitude and frequency modulation of vibro-acoustics [76]. However, most of these studies address breathing phenomenon in a through crack whereas the methodology to detect part-through breathing crack in a plate has yet to be researched.

2.4.3 Random Excitation

Random excitation has been used by various researchers for crack detection. The non-linear indicator functions for random signals includes the bispectral analysis [77]. The third order spectrums have been used by measuring skewness to detect [15] and locate crack numerically [78]. Strain-time history graphs have been also used to find the location and depth of breathing cracks by using Continuous Wavelet Transform (CWT) [79]. Detection of crack is quite accurate but for severity, difference in numerical and experimental results has been observed. The study has been extended for comparison with open and breathing cracks in a freely suspended beam presenting vehicle-bridge model and it has been shown that CWT shows better result for breathing crack than for open crack [80]. However, these studies have utilized the baseline data and do not give a clear severity estimation. From the above mentioned studies it has been observed that although various excitations have been explored discretely for crack identification and quantification in case of beams but a combined comparison analysis for part-through breathing cracks in a plate-like structure under different excitations have to be investigated yet.

2.5 SIGNAL PROCESSING TECHNIQUES

Different signal processing techniques can be applied to the response signal of the damaged structure. The processing techniques are broadly classified into three i.e. time domain, frequency domain and time-frequency domain.

2.5.1 Time Domain

The time domain models are based on statistical time series models. The probability distribution and density functions are used for the detection of crack by finding skewness and kurtosis. The skewness is the third moment of the power spectrum. It indicates the asymmetry in the power density function. The crack identification using the time series data was initiated in early 2000's. Pattern recognition techniques like principal component analysis, neural networks uses the time series data for crack detection. In 2006, the Hidden Markov Model for early detection of crack was developed by incorporating partitioning to the time series data [81]. In 2008, the crack was investigated using the statistical distribution of state space points on the attractor and comparing it with Poincare′ map [82]. The study revealed that the skewness provided better localization as compared to the variance based method. Another breakthrough in evaluating time series data was the empirical mode decomposition (EMD). In 2010, crack was identified by EMD and quantified in the form of energy damage index. The study revealed that as the crack depth ratio increased, the value of the damage index became greater [83]. Another researcher used the entropy measurement as the criteria to measure nonlinearity induced in the system due to the presence of crack [84, 85]. The phase diagrams were also used to identify the nonlinearity of cracks due to breathing [86]. Recently, the breathing crack has also been identified and localized using a forward and an inverse problem in a shear structure [87]. These studies were conducted on cracked beams of various severities. In a plate, the crack identification and localization was performed using Bayesian parameter estimation in 2011 [88]. The Lamb waves were also used for crack identification and quantification using multivariate regression model [89] and for localization by defining a damage index (DI) based on the Time of flight between sensors and then forming an image by image fusion scheme based on DI [90].

2.5.2 Frequency Domain

The Fast Fourier transform (FFT) has been widely used for the detection of damage by the measurement of change in natural frequency of the structures. FFT can be used effectively for linear responses but for nonlinear responses, it is unable to characterize damage effectively. Although FFT is a fair indicator of damage by showing harmonics in the spectrum yet advanced signal processing techniques need to be applied to locate and estimate damage. It gives change in frequency domain only and ignores the frequency varying over time.

In the past two decades, the breathing crack has been explored extensively by using FFT spectrum. The superharmonics produced due to breathing of crack were investigated by varying force [19], by applying Singular Value Decomposition (SVD) technique [91] and enhanced weighting function on Fourier power spectrum [92]. The higher harmonics and intermodulations were also used to detect multi-cracks [93]. In 2015, another researcher compared the frequency- and time-displacement spectrums of an open crack model with that of a breathing crack model of a beam [60]. It was found that greater the crack depth, more would be the amplitude of harmonics. The FFT was advanced by introducing a method of wave mixing for detection of breathing cracks. The low frequency wave was mixed with a high frequency wave and interacted with the damaged beam. The low frequency wave excited the crack to open and close while the high frequency wave localized the damage [70]. The wave mixing method was extended to a plate as well [64]. Another researcher, studied the change in natural frequency with respect to crack depth and angle of an oblique breathing crack in a beam [72]. Single and bi-tone has been used in another research for exciting the structure to get the response from the structure. Low energy components from the higher order harmonics were extracted by using a weighting function on the spatial curvature for damage detection of breathing crack. [94]

All these methods have been used on beam-like structures while plate-like structures having breathing cracks still needs to be explored.

2.5.3 Time-Frequency Domain

The time-frequency domain includes HT, HHT, Wavelet transform, Wigner-Ville Transforms and Short Time Fourier Transform.

The HT gives the frequencies of the response over time. In case of nonlinear dynamical systems such as breathing cracks, HT has been found to be effective. In 2005, a researcher investigated the dynamic behavior of the breathing crack in a beam by using Instantaneous frequencies. It also revealed the variation of Instantaneous frequencies upon opening and closing of the crack depends on the crack depth and introduced bilinear frequencies for breathing crack [24]. Another researcher, experimented with a plate having through crack and used HHT to find the instantaneous amplitude and instantaneous frequency as an indicator of damage [76]. In 2013, a novel method using HT was used to detect breathing crack by using piecewise equations [57]. It revealed that during the positive half wave, the crack opens and during the negative half wave, the crack closes, hence, representing the intact beam frequency. A formula was derived to find out the bilinear frequency from these two frequencies.

Another research focuses on finding the bilinear amplitude and bilinear frequency approximations of a beam having a breathing crack under prestress or gap [95]. The study on breathing crack in a beam was enhanced to multi crack detection and localization by another researcher by using HHT [62]. Another research uses a slow flow model based on complexicating-averaging technique and HT to detect breathing crack [96]. Research shows that the wavelet technique was also used to locate and estimate breathing crack in a beam [79]. In 2017, Wavelet Transform was also used to detect multi-breathing cracks in a beam. Breathing crack was also identified by using statistical entropy combined with wavelet technique [84].

The common modeling approach for breathing cracks in plate incorporated the Contact Acoustic Nonlinearity (CAN) in the bilinear model, by interacting the Guided Ultrasonic Waves (GUW) on the through-thickness crack [97] while some studies used the Coulomb friction model at the crack interface and analyzed its nonlinear scattering and mode conversion [98]. These studies were limited to very thin plates with through cracks. However, an all-over part-through crack in a thin plate was modeled analytically and the bifurcation study performed [3]. Later on, the bifurcation study was extended to thick plate [99]. Another study, used the bilinear stiffness model for breathing crack in a beam and plate and compared the effect of geometric nonlinearity between the intact and damaged structure [100].

The above review suggests that the detection of breathing crack in both beam and plate type structures has been done under different excitations i.e. harmonic, impulse and random along with different techniques. However, the crack location and severity in 2D structures have been touched in some cases with respect to different parameters like natural frequency and second order harmonics but further research in this area is still required. Moreover, most techniques mentioned above are dependent on the baseline data of the structure. The crack presence affects both macro- and micro-structures hence, efficient damage detection is needed. The environmental conditions can affect baseline data, hence generating false indications of damage. Thus, a robust technique needs to be formulized that can do accurate identification and severity estimation of crack on 2D structures without depending on baseline-data.

2.6 NONLINEAR SOLUTION METHODOLOGIES

The solution methodologies maybe classified into two groups, exact analytical solution or approximate analytical solution. The solution applicable to the study of dynamic behavior of nonlinear deflection of plate are the approximate analytical solutions. It is evident from literature that the solution for differential equation for a nonlinear case is complicated as compared to linear problems. Nonlinear methods include finite difference technique, double Fourier series, Galerkin method, Berger formulation, Hamilton's principle, Volterra series, Rayleigh-Ritz technique etc.

2.6.1 Perturbation Theory

Perturbation theory is for weakly nonlinear problems. Many methods exist for the perturbation theory like method of multiple scales, incremental harmonic balance, averaging, krylov-bogolioubov, Lindstedt Poincare method. Nayfeh was the first to use the perturbation method of multiple scale to a symmetric plate. Till date perturbation theory is being utilized by combining with different techniques to solve nonlinear equation of cracked rectangular plates like Krylov-bogoliubov method [101], Hamilton's principle [102] and Linstedt Poincare technique [103].

The perturbation method was first applied for detecting a breathing crack in a beam for various crack stiffness and depth parameters in 2010. It also compared breathing crack model to that of open crack model with the experimental results and found close agreement to the breathing crack model [68]. A step further was taken towards breathing crack detection by nonlinear methods, by applying a polynomial of higher order and the results verified experimentally [55]. A tapered beam with a double edge breathing crack was investigated by Rayliegh-Ritz method in another study [104]. This study showed that the response amplitude varied with the excitation level increasing with crack severity. The breathing crack in a beam was studied in another research by estimating the adaptive Volterra filter kernels [93].

2.7   RESEARCH GAP

The literature review gives us insight on the vast exploration in the field of SHM according to the problems posed by the real-life structures. The review not only highlights different plate theories and nonlinear solution methodologies but also discusses the advancement in signal processing techniques to help in damage detection. This research focuses on the thin plate theory and application of the method of multiple scales on cracked rectangular plates. The cubic nonlinear equation is then written in piecewise equations form to represent the breathing crack dynamics. This is a major contribution towards understanding and simplifying the nonlinear systems.

3.MATHEMATICAL MODEL OF A BREATHING CRACK

The uncertainties and assumptions in the linear open crack model led to the research of real-life breathing crack nonlinear analysis in a plate. The appearance of cracks in structures made of light weight and flexible material is of prime interest in modern technology. Therefore, the presented research is focused on the detection of cracks in light weight thin aluminum plates. Rectangular plates are used as a basic building block of many structures. Hence, a rectangular plate is selected.

3.1   INTRODUCTION

Many mathematical models for a breathing crack in beams have been developed in the last two decades. However, this thesis models a breathing crack in a plate. For this purpose, the thin plate theory i.e. Kirchoff-Love plate theory has been considered. The assumptions used in the plate modeling are:

1. Material of the plate is isotropic, perfectly elastic and homogeneous with uniform thickness.

2. Hooke's law is applicable.

3. Middle axis of the plate remains same after deflection

Under vibration, the rectangular plate with length l, breadth b and thickness h in x-, y- and z- directions, respectively and D is the flexural rigidity $= \frac{Eh^3}{12(1-v^2)}$, E is the modulus of elasticity, v is the Poisson's ratio, ρ is the density, undergoes lateral deflection

$w(x, y, t)$, under a force of $P_z(x, y, t)$ per unit area acting on the surface. It uses the differential equation (3.1) as:

$$\nabla^4 w + \rho \frac{\partial^2 w}{\partial t^2} = P_z \tag{3.1}$$

If the plate has a crack in the center of length $2a$ which is parallel to the length l of the plate, the Sophie-Germaine equation for the transverse deflection w of plate can be modified for plate having part-through crack in the middle as shown by equation (3.2) [5]:

$$D[\frac{\partial^4 w}{\partial x^4} + 2\frac{\partial^4 w}{\partial x^2 \partial y^2} + \frac{\partial^4 w}{\partial y^4}] = -\rho h \frac{\partial^2 w}{\partial x^2} + P_z(x, y, t) + n_x \frac{\partial^2 w}{\partial x^2} +$$
$$\frac{2aD(\frac{\partial^4 w}{\partial y^4} + v\frac{\partial^4 w}{\partial x^2 \partial y^2})}{3(\alpha_{bt}^0/6 + \alpha_{bb}^0)(3+v)(1-v)h + 2a} - \frac{2a}{(6\alpha_{bt}^0 + \alpha_{tt}^0)(1-v^2)h + 2a} n_{rs} \frac{\partial^2 w}{\partial y^2} \tag{3.2}$$

In equation (3.2), α_{bb}^0 is the bending compliance coefficient and α_{bt}^0 is the bending-tensile compliance coefficient, n_x is in-plane force in the x-direction, n_{rs} is in-plane force at far side of the plate, α_{bb} is bending compliance at crack center, α_{bt} is stretching-bending compliance at crack center and α_{tt} is stretching compliance at crack center. The equation (3.2) can be written in the form of the equation of motion with cubic nonlinearity by using Galerkin method [105]. Let the solution be written as equation (3.3):

$$w(x, y, t) = \sum_{n=1}^{\infty} \sum_{m=1}^{\infty} A_{mn} X_m Y_n U_{mn}[t] \tag{3.3}$$

where, A_{mn} refers to the arbitrary amplitude, X_m, Y_n are the modal functions and $U_{mn}[t]$ is the time dependent modal coordinate. To find the in-plane forces n_x and n_{rs}, Berger formulation is applied by equation (3.4) and (3.5), respectively.

$$n_x = \frac{Eh}{(1-v^2)}(\varepsilon_x + v\varepsilon_y) \tag{3.4}$$

$$n_{rs} = \frac{Eh}{(1-v^2)}(\varepsilon_y + v\varepsilon_x) \tag{3.5}$$

The middle surface strains ε_x and ε_y are derived by Timoshenko are defined by equations (3.6)and (3.7), respectively.

$$\varepsilon_x = \frac{\partial u}{\partial x} + \frac{1}{2}(\frac{\partial w}{\partial x})^2 \tag{3.6}$$

$$\varepsilon_y = \frac{\partial v}{\partial y} + \frac{1}{2}(\frac{\partial w}{\partial y})^2 \tag{3.7}$$

Therefore, substituting the values ε_x and ε_y from equations (3.6) and (3.7) in equations (3.4) and (3.5), we get equations (3.8) and (3.9),

$$\frac{n_x h^2}{12D} = \frac{\partial u}{\partial x} + v\frac{\partial v}{\partial y} + \frac{1}{2}(\frac{\partial w}{\partial x})^2 + \frac{1}{2}v(\frac{\partial w}{\partial y})^2 \tag{3.8}$$

$$\frac{n_{rs} h^2}{12D} = \frac{\partial v}{\partial y} + v\frac{\partial u}{\partial x} + \frac{1}{2}(\frac{\partial w}{\partial y})^2 + \frac{1}{2}v(\frac{\partial w}{\partial x})^2 \tag{3.9}$$

By integrating over the length in x and y direction, multiplying by $dxdy$ and neglecting u and v at the external BCs, it can be written as equations (3.10) and (3.11)

$$\frac{n_x h^2 lb}{12D} = \frac{1}{2}\int_0^l \int_0^b [(\frac{\partial w}{\partial x})^2 + \frac{1}{2}v(\frac{\partial w}{\partial y})^2]dxdy \tag{3.10}$$

$$\frac{n_{rs} h^2 lb}{12D} = \frac{1}{2}\int_0^l \int_0^b [(\frac{\partial w}{\partial y})^2 + \frac{1}{2}v(\frac{\partial w}{\partial x})^2]dxdy \tag{3.11}$$

Rearranging the equations (3.10) and (3.11), n_x and n_{rs} can be written as equations (3.12) and (3.13)

$$n_x = DC_{1mn}A_{mn}^2 U_{mn}^2(t) \tag{3.12}$$

$$n_{rs} = DC_{2mn}A_{mn}^2 U_{mn}^2(t) \tag{3.13}$$

where, A_{mn} is the modal peak amplitude frequency normalized to unity.

$$C_{1mn} = \frac{6}{lbh^2} \int_0^l \int_0^b [X_m'^2 Y_n^2 + v Y_n'^2 X_m^2] dxdy \qquad (3.14)$$

$$C_{2mn} = \frac{6}{lbh^2} \int_0^l \int_0^b [Y_n'^2 X_m^2 + v X_m'^2 Y_n^2] dxdy \qquad (3.15)$$

Replacing $w(x, y, t)$ and n_x and n_{rs} in equation (3.2), the final form is obtained as a cubic nonlinear equation as (3.16).

$$M_{mn}\ddot{U}_{mn}[t] + K_{mn}U_{mn}[t] + G_{mn}U_{mn}^3[t] = P_{mn}[t] \qquad (3.16)$$

where, M_{mn} is the mass, K_{mn} is the stiffness, G_{mn} is the co-efficient of the cubic non-linear term, P_{mn} is the excitation. Let, $m = n = 1$ for first mode which can then be calculated by the equations (3.17) to (3.19).

$$M_{mn} = \frac{\rho h}{D} \sum_{m=1}^{\infty} \sum_{n=1}^{\infty} A_{mn} \int_0^l \int_0^b X_m^2 Y_n^2 dxdy \qquad (3.17)$$

$$K_{mn} = \sum_{m=1}^{\infty} \sum_{n=1}^{\infty} A_{mn} \int_0^l \int_0^b (X_m^{iv} Y_n + 2X_m'' Y_n'' + Y_n^{iv} X_m - \frac{2a(v X_m'' Y_n'' + Y_n^{iv} X_m)}{3(\alpha_{bt}^0/6 + \alpha_{bb}^0)(3 + v)(1 - v)h + 2a}) X_m Y_n dxdy \qquad (3.18)$$

$$G_{mn} = \sum_{m=1}^{\infty} \sum_{n=1}^{\infty} A_{mn}^3 \left(-C_{1mn} X_m X_m'' Y_n^2 + \frac{2a(C_{2mn} X_m^2 Y_n Y_n'')}{(6\alpha_{bt}^0 + \alpha_{tt}^0)(1 - v^2)h + 2a} \right) dxdy \qquad (3.19)$$

$$P_{mn}(t) = P_0(t) X_m(x_0) Y_n(y_0) \qquad (3.20)$$

where, $P_0(t)$ = Point force and $X_m(x_0) Y_n(y_0)$ = Location of force.

The modal functions X_m and Y_n depend upon various BCs. By substituting the values of X_m and Y_n, values of mass, stiffness and cubic non-linearity can be calculated. This thesis discusses four different BCs with their modal functions as given in Table 3.1:

Table 3.1: X_m and Y_n Formulas for Various BCs

All Sides Simply Supported (SSSS)
$X_m = sin\frac{\pi x}{l}$ $Y_n = sin\frac{\pi y}{b}$
Clamped-Clamped-Simple-Simple (CCSS)
$X_m = sin\frac{\pi x}{l}sin\frac{\pi x}{2l}$ $Y_n = sin\frac{\pi y}{b}sin\frac{\pi y}{2b}$
Clamped-Clamped-Free-Free (CCFF)
$X_m = cos\frac{\lambda_m x}{l} - cosh\frac{\lambda_m x}{l} - \gamma\left(sin\frac{\lambda_m x}{l} - sinh\frac{\lambda_m x}{l}\right)$ $Y_n = cos\frac{\lambda_n y}{b} - cosh\frac{\lambda_n y}{b} - \gamma\left(sin\frac{\lambda_n y}{b} - sinh\frac{\lambda_n y}{b}\right)$
Clamped-Free-Free-Free (CFFF)
$X_m = 1$ $Y_n = cos\frac{\lambda_n y}{b} - cosh\frac{\lambda_n y}{b} - \gamma\left(sin\frac{\lambda_n y}{b} - sinh\frac{\lambda_n y}{b}\right)$

In most engineering applications, the first mode has the greatest impact so mode I has been considered in further equations [106]. The equation (3.16) known as Duffing equation can be rewritten as:

$$\ddot{U}_{mn}[t] + \omega^2 U_{mn}[t] + \epsilon U_{mn}^3[t] = \gamma_{mn}[t] \tag{3.21}$$

$$\omega_{mn}^2 = \frac{K_{mn}}{M_{mn}}, \epsilon_{mn} = \frac{G_{mn}}{M_{mn}}, \gamma_{mn} = \frac{P_{mn}}{M_{mn}}$$

ω_{mn} = natural frequency of the cracked rectangular plate,

ϵ_{mn} = co-efficient of the nonlinear cubic term,

γ_{mn} = the input force to mass matrix ratio.

The value of the ϵ_{mn} can be numerically positive or negative which shows that the response may show increased or decreased stiffness.

3.2 BILINEAR SYSTEM DYNAMIC CHARACTERISTICS

In this thesis, the developed equation (3.21) will be further used for calculating the bilinear frequencies of the various crack length for various BCs. Let the plate with central crack undergo free transverse vibration ($P_{mn}[t] = 0$) which causes the crack to open and close. From the literature, it is evident that on hogging the crack is opened (representing open crack plate frequency) and on sagging the crack is closed (representing intact plate frequency) [107] as shown in Figure 3.1. During the crack's open state, the frequency is ω_o and during its closed state, the frequency changes to ω_c, ω_c being the intact plate frequency. ϵ_1 and ϵ_2 are the open and closed nonlinear terms, respectively. During the closed state ϵ_2 can be supposed to be approaching to 0.

Hence, a set of bilinear equations from the basic equation has been obtained in this re-

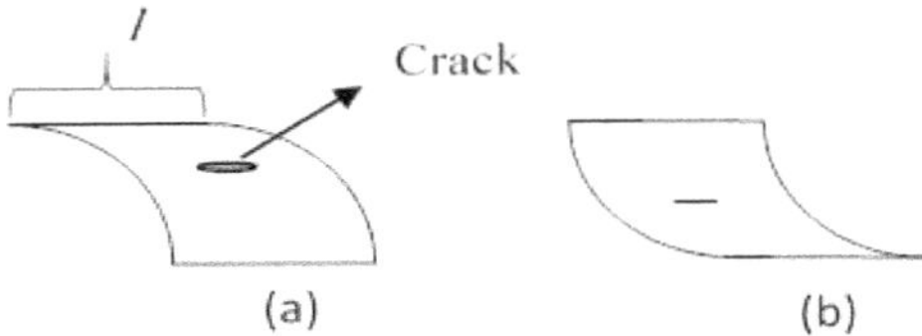

Figure 3.1: Breathing of crack in a plate (a) Open crack (hogging) (b) Close crack (sagging)

search which is written in the piece-wise form as shown by equation (3.22) and (3.23).

$$\ddot{U}_{mn}[t] + \omega_o^2 U_{mn}[t] + \epsilon_1 U_{mn}^3[t] = 0; U_{mn} \geq 0 \tag{3.22}$$

$$\ddot{U}_{mn}[t] + \omega_c^2 U_{mn}[t] + \epsilon_2 U_{mn}^3[t] = 0; U_{mn} \leq 0 \tag{3.23}$$

As in real-time, the system will be damped so if the damping effect μ is considered, the equations can be written as equations (3.24) and (3.25):

$$\ddot{U}_{mn}[t] + 2\mu\dot{U}_{mn}[t] + \omega_o^2 U_{mn}[t] + \epsilon_1 U_{mn}^3[t] = 0; U_{mn} \geq 0 \tag{3.24}$$

$$\ddot{U}_{mn}[t] + 2\mu\dot{U}_{mn}[t] + \omega_c^2 U_{mn}[t] + \epsilon_2 U_{mn}^3[t] = 0; U_{mn} \leq 0 \tag{3.25}$$

Consider that, during the positive half cycle the crack opens and during the negative half cycle, the crack closes. The breathing crack behaves as a bilinear system in which the stiffness varies each time the oscillator passes the stiffness interface $U[t] = 0$. The stiffness of the open crack and the closed crack combines to form the stiffness of the breathing crack $k(u)$. So, it can be said that the positive half region of the response gives the open crack stiffness, k_o and the negative half region of the response gives the closed crack stiffness, k_c as shown by equation (3.26)

$$k(u) = \begin{cases} k_o, u > 0 \\ \\ k_c, u \le 0 \end{cases} \tag{3.26}$$

It is expressed by equation (3.27) derived from equation (3.18) by taking $2a = 0$. Equation (3.18) represents the open crack stiffness i.e. $k_o = K_{mn}$.

$$k_c = \sum_{m=1}^{\infty} \sum_{n=1}^{\infty} A_{mn} \int_0^l \int_0^b (X_m^{iv} Y_n + 2X_m'' Y_n'' + Y_n^{iv} X_m) X_m Y_n dx dy \tag{3.27}$$

The bilinear equations give two frequencies for two regions. Therefore, two stiffness parameters k_o and k_c exist. The reduced stiffness is given as, $k_o = \alpha k_c$ where α is the stiffness ratio between 0 and 1. When the crack is closed $\alpha = 1$ and when it is open $0 \le \alpha \le 1$. The expression for finding ω_o and ω_c are $\omega_o = \sqrt{\frac{k_o}{M}}$ and $\omega_c = \sqrt{\frac{k_c}{M}}$, respectively. These frequencies can then be used to find the breathing crack frequency ω_b as,

$$\omega_b = \frac{2\omega_c\omega_o}{\omega_c + \omega_o} \tag{3.28}$$

The difference in frequencies in the two stiffness regions is $\Delta\omega$ and the direct difference in natural frequencies between intact and cracked plate is $\Delta\omega_{br}$. They can be calculated by equation (3.29) and (3.30).

$$\Delta\omega = \omega_c - \omega_o \tag{3.29}$$

$$\Delta\omega_{br} = \omega_c - \omega_b \tag{3.30}$$

The significance of finding the positive half and negative half frequencies separately is that their difference can identify the breathing crack. This method considers the dynamic characteristics of bilinear systems for crack identification.

Substituting the values of X_m and Y_n of the BCs in equations (3.17) to (3.19), the frequencies for the first mode are obtained analytically. $Mathematica^{TM}$ has been used for the calculation of the frequencies by the formulas of ω_o and ω_c . The crack severity ratio (crack length/length of plate = $2a/l$) has been used for identifying various crack lengths. The calculations has been done on a plate $(0.5 \times 1 \times 0.01\,m)$, with material properties $E = 70.3\,GPa$, and $\rho = 2660\,kg/m^3$. The crack depth, $d = 0.005\,m$ is the same for all cases. The calculated results are shown in Table 3.2. Figure 3.2 highlights the drop in natural frequencies for various crack severity ratios under different BCs. The difference in natural frequencies, $\Delta\omega_{br}$ calculated directly between the breathing cracked plates, ω_b and intact plate with different severities show lesser difference as compared to the proposed method in equation (3.29).

$$\Delta\omega \geq \Delta\omega_{br} \tag{3.31}$$

Table 3.2: Analytical values of 1st natural frequency (Hz) of plate with open and breathing crack with various BCs

BCs	Freq. (Hz)	Crack severity ratio ($2a/l$)						
		Intact	0.05	0.1	0.2	0.3	0.4	0.5
CCFF	ω_o	51.6744	51.2184	50.8828	50.4221	50.1207	49.9081	49.7501
	ω_{br}	51.6744	51.4469	51.2802	51.0521	50.9035	50.7989	50.7214
CCSS	ω_o	178.803	177.671	176.84	175.702	174.959	174.435	174.047
	ω_{br}	178.803	178.238	177.824	177.259	176.891	176.633	176.441
SSSS	ω_o	123.473	122.461	121.747	120.806	120.213	119.805	119.508
	ω_{br}	123.473	123.03	122.675	122.208	121.916	121.715	121.568
CFFF	ω_o	8.8008	8.08518	7.51978	6.67288	6.05985	5.58968	5.21433
	ω_{br}	8.8008	8.4506	8.18543	7.8097	7.5557	7.37224	7.23341

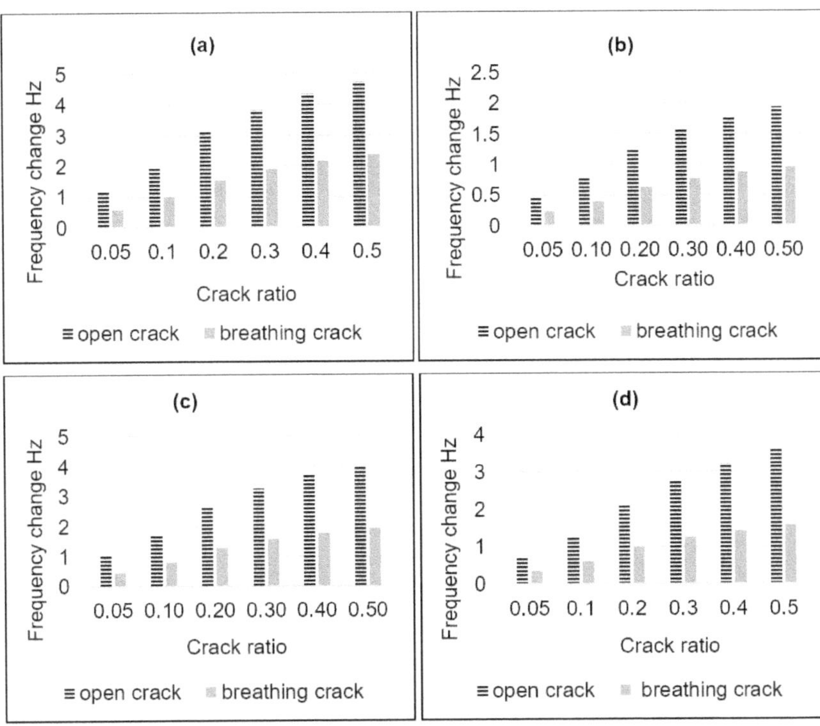

Figure 3.2: Frequency change between open cracks and breathing cracks relative to intact plate (a) CCFF (b) CCSS (c) SSSS (d) CFFF

It can be observed from Figure 3.2 that only minor decrease in frequency occurs in breathing cracks as compared to open cracks, as expressed by equation (3.31). Hence, breathing cracks cannot be identified with conventional methods presented for open cracks. For this reason, the breathing behavior of the crack has been observed and modeled to indicate damage in structure.

3.3   VALIDATION OF THE DEVELOPED MATHEMATICAL MODEL

The developed mathematical model of a plate having a breathing crack has been validated by comparing it with the published research of all-over breathing crack for plate
and beam. The central crack has been extended to an all-over crack for validation purpose. The comparison has been made for plates with different boundary conditions.
As an analytical model for central part-through breathing crack in a plate has not been
published yet. Therefore, the comparison has been made with the open part-through
crack for CCSS plate ($0.5 \times 1 \times 0.01$ m), with material properties ($E = 70.3$ GPa and ρ
$= 2660$ kg/m^3) and results show that error is lesser for the presented method [5]. The
method has been validated by using the same specifications to an SSSS plate ($1 \times 2 \times$
0.008 m), with material properties ($E = 210$ GPa and $\rho = 7860$ kg/m^3) and centrally
located crack (all-over, part-through crack) as in the published literature [3]. Similarly,
the analytical method of a breathing crack for cantilever case ($0.015 \times 1 \times 0.05$ m), ($E =$
206 GPa and $\rho = 7850$ kg/m^3) has been compared by applying the present study [57].
An all-over breathing crack simulated plate model ($0.07 \times 0.706 \times 0.002$ m), with material properties ($E = 73$ GPa and $\rho = 2850$ kg/m^3), has been validated for a cantilever
case by comparing with a recent publication [100]. The comparison shows that the
values of intact and cracked case are same, hence validating the simulated model. The
developed simulated model has been thus used as reference for analytical results.

Table 3.3: Comparison of 1st natural frequency (Hz) of plate with part-through breathing crack between published and proposed work

Validation studies	BCs	1st Natural Frequency (Hz)		Error % between Published and Proposed work	
		Intact plate	Cracked plate	Intact plate	Cracked plate
Civera [2019] - Simulated [100]	CFFF	3.310	3.307	0	0
Proposed work	CFFF	3.310	3.307		
Chen [2016] - Analytical [3]	SSSS	23.9874	23.8698	3.554	1.964
Proposed work	SSSS	24.84	24.348		
Chen [2016] - Simulated [3]	SSSS	25.101	24.941	1.250	0.633
Proposed work	SSSS	24.787	24.783		
Yan [2013] - Analytical [57]	CFFF	43.00	39.12	0.116	7.31
Proposed work	CFFF	43.05	41.98		
Israr [2008] - Analytical [5]	CCSS	184.99	183.800	3.344	3.25
Proposed work	CCSS	178.803	177.824		

Table 3.3 shows the comparison of proposed method and the values of frequency from literature. It is noted that the results of the proposed method agree well with the published work i.e. the error is less than 10%. The cantilever has the most error so by considering the worst case scenario, the methodology would be validated and discussed in further detail.

3.4   SOLUTION METHODS OF NONLINEAR EQUATIONS

The set of nonlinear equations developed in section 3.2 can be solved by a number of analytical methods. The Method of Multiple Scales have been used in this research to provide the response equation. Free and forced inputs are used to derive the response equations.

3.4.1   Impulse Excitation

For the derivation of impulse excitation response, it is assumed that the system is initially disturbed and then freed. So the derivation will be proceeded from equation (3.24) and (3.25). In this method, it is assumed that the system is perturbed in zeroth and first order ($\varepsilon^n, n = 0, 1, 2, \ldots$) and time varies from fast to slow such that,

$T_0 = \varepsilon^0 t, T_1 = \varepsilon^1 t$

Let the initial conditions be,

$U(0) = \lambda_o, \dot{U}(0) = 0$

where, λ_o is the initial displacement amplitude. The zeroth and the first order of approximation is inserted in equation (3.24) and separated as equations (3.32) and (3.33), respectively.

$$O(\varepsilon^o) : D_o^2 U_o + \omega^2 U_o = 0 \tag{3.32}$$

$$O(\varepsilon^1) : D_o^2 + U_1 = -2D_o D_1 U_o - 2\mu D_o U_o - \epsilon U_o^3 \tag{3.33}$$

The solution at the zeroth order can be written in complex exponential form as shown by equation (3.34)

$$U_o = \Gamma e^{i\omega T_o} + \bar{\Gamma} e^{i\omega T_o} \tag{3.34}$$

where, Γ is the unknown complex amplitude and $\bar{\Gamma}$ is its complex conjugate. For solving, it can be written in the polar form as equation (3.35)

$$\Gamma = \frac{1}{2}\lambda e^{i\alpha} \tag{3.35}$$

where, λ is the real amplitude and α is the real phase w.r.t T_1.

The first order solution as shown in equation (3.36), is obtained by substituting equation (3.34) in equation (3.33).

$$D_o^2 U_1 + \omega^2 U_1 = -2D_o D_1 (\Gamma e^{i\omega T_o} + \bar{\Gamma} e^{-i\omega T_o}) - 2\mu D_o (\Gamma e^{i\omega T_o} + \bar{\Gamma} e^{-i\omega T_o})$$
$$-\epsilon(\Gamma e^{i\omega T_o} + \bar{\Gamma} e^{-i\omega T_o})^3 \tag{3.36}$$

$$D_o^2 U_1 + \omega^2 U_1 = -2i\omega D_1 \Gamma e^{i\omega T_o} + 2i\omega D_1 \bar{\Gamma} e^{-i\omega T_o} - 2i\omega\mu\Gamma e^{i\omega T_o}$$
$$+2i\omega\mu\bar{\Gamma} e^{-i\omega T_o} - \epsilon\Gamma^3 e^{3i\omega T_o} - 3\epsilon\Gamma^2\bar{\Gamma} e^{i\omega T_o} - 3\epsilon\Gamma\bar{\Gamma}^2 e^{-i\omega T_o} - \epsilon\Gamma^3 e^{-3i\omega T_o} \tag{3.37}$$

The secular terms as expressed by equation (3.38), that cause non-uniformity can be removed and by placing the following terms equal to zero as,

$$(-2i\omega D_1\Gamma - 2i\omega\mu\Gamma - 3\epsilon\Gamma^2\bar{\Gamma})e^{i\omega T_o} = 0 \tag{3.38}$$

After substitution and elimination of secular terms, equation (3.37) can be rewritten as

$$D_o^2 U_1 + \omega^2 U_1 = -\epsilon\Gamma e^{3i\omega T_o} - \epsilon\bar{\Gamma} e^{-3i\omega T_o} \tag{3.39}$$

Converting equation (3.39) into complex conjugate (cc) form as

$$D_o^2 U_1 + \omega^2 U_1 = -\epsilon\Gamma e^{3i\omega T_o} + cc \tag{3.40}$$

The solution can then be written by assuming the response in the form of equation (3.41) and then by double differentiating it, equation (3.42) can be obtained.

$$U_1 = A_1 e^{3i\omega T_o} + A_2 e^{-3i\omega T_o}$$ (3.41)

$$\ddot{U}_1 = -9\omega^2 A_1 e^{3i\omega T_o} - 9\omega^2 A_2 e^{-3i\omega T_o}$$ (3.42)

Substituting equation (3.41) and (3.42) in equation (3.40) and solving for A_1 and A_2 gives:

$$A_1 = \frac{\epsilon}{8\omega^2}\Gamma^3, \ A_2 = \frac{\epsilon}{8\omega^2}\bar{\Gamma}^3$$

Hence, the solution at first order is

$$U_{0,1} = \Gamma e^{i\omega T_o} + \frac{\epsilon}{8\omega^2}\Gamma^3 e^{3i\omega T_o} + cc$$ (3.43)

In terms of trigonometric identities, solution becomes

$$U(\varepsilon^0 t, \varepsilon^1 t_1) = \lambda cos(\omega t - \alpha) + \frac{\epsilon}{32\omega^2}\lambda^3 cos(3(\omega t - \alpha))$$ (3.44)

For the crack to be closed ϵ becomes negligible and equation (3.44) reduces to $\lambda cos(\omega t - \alpha)$. Hence, the set of bilinear response equations are obtained as,

$$U[t] = \lambda cos(\omega t - \alpha) + \frac{\epsilon}{32\omega^2}\lambda^3 cos(3(\omega t - \alpha)); U \geq 0$$ (3.45)

$$U[t] = \lambda cos(\omega t - \alpha); U \leq 0$$ (3.46)

The equations (3.45) and (3.46) will be used to obtain the response of cracked plate during opening and closing. The physical model of the breathing crack response cycle becomes asymmetric with the time series by allocating positive half cycle to the opening of crack and negative half cycle to the closing of crack. By separating the response into positive and negative halves and then taking the HT of both halves, the corresponding Instantaneous Frequencies (IF) can be obtained and compared.

3.4.2 Harmonic Excitation

The derivation of the response as mentioned in previous section has been formulated for impulse excitation i.e. when force $P_{mn} = 0$. However, during working condition, the structure is under some excitation. Let the system excited by a harmonic force of $A\cos\omega t$. Then the input can be written in equation (3.24) and (3.25) can be shown by equation (3.47) and (3.48) as:

$$\ddot{U}_{mn}[t] + 2\mu\dot{U}_{mn}[t] + \omega_o^2 U_{mn}[t] + \epsilon_1 U_{mn}^3[t] = A\cos\omega t; U_{mn} \geq 0 \qquad (3.47)$$

$$\ddot{U}_{mn}[t] + 2\mu\dot{U}_{mn}[t] + \omega_c^2 U_{mn}[t] + \epsilon_2 U_{mn}^3[t] = A\cos\omega t; U_{mn} \leq 0 \qquad (3.48)$$

The zeroth and the first order of approximation is inserted in equation (3.47) and separated as equations (3.49) and (3.50), respectively,

$$O(\epsilon^o) : D_o^2 U_o + \omega^2 U_o = 0 \qquad (3.49)$$

$$O(\epsilon^1) : D_o^2 + U_1 = -2D_o D_1 U_o - 2\mu D_o U_o - \epsilon U_o^3 + A\cos\omega t \qquad (3.50)$$

As before, the solution is written in complex exponential form as shown by equation (3.51)

$$U_o = \Gamma e^{i\omega T_o} + \bar{\Gamma} e^{i\omega T_o} \tag{3.51}$$

Also, $A\cos\omega t$ can be written in exponential form after applying $\Omega = \omega + \varepsilon\sigma$ as perturbation

$$A\cos\Omega t = \frac{A}{2}e^{j\Omega t} + \frac{A}{2}e^{-j\Omega t}$$

This is the primary resonance condition where σ is the detuning factor and Ω is the excitation frequency which is close to the natural frequency, ω. Putting the complex exponential values in equation (3.50), one gets the equation (3.52)

$$D_0^2 U_1 + \omega^2 U_1 = e^{i\omega T_o}(-2i\omega D_1\Gamma + 2i\omega D_1\bar{\Gamma}e^{-2i\omega T_o} - 2i\hat{\mu}\omega\Gamma + 2i\hat{\mu}\omega\bar{\Gamma}e^{-2i\omega T_o}$$
$$-\hat{\epsilon}\Gamma^3 e^{2i\omega T_o} - 3\hat{\epsilon}\Gamma^2\bar{\Gamma} - 3\hat{\epsilon}\Gamma\bar{\Gamma}^2 e^{-2i\omega T_o} - \hat{\epsilon}\bar{\Gamma}^3 e^{-4i\omega T_o} \tag{3.52}$$
$$+\frac{A}{2}e^{i\varepsilon\sigma T_o} + \frac{A}{2}e^{-i2(\omega+\varepsilon\sigma)T_o})$$

By removing the secular terms i.e. equation (3.53),

$$(-2i\omega D_1\Gamma - 2i\omega\mu\Gamma - 3\epsilon\Gamma^2\bar{\Gamma} + \frac{A}{2}e^{i\varepsilon\sigma T_o})e^{i\omega T_o} = 0 \tag{3.53}$$

The first order equation with secular terms removed will be of the form of equation (3.54)

$$D_0^2 U_1 + \omega^2 U_1 = e^{i\omega T_o}(-\hat{\epsilon}\Gamma^3 e^{2i\omega T_o} - \hat{\epsilon}\bar{\Gamma}^3 e^{-4i\omega T_o}) \tag{3.54}$$

Multiplying the right hand side of equation (3.54), the equation (3.55) is obtained.

$$D_0^2 U_1 + \omega^2 U_1 = -\hat{\epsilon}\Gamma^3 e^{3i\omega T_o} - \hat{\epsilon}\bar{\Gamma}^3 e^{-3i\omega T_o} \tag{3.55}$$

This acquires the same form as equation (3.39). Hence, the solution takes the form as equation (3.45) and (3.46). However, the natural frequency would be replaced by the excitation frequency. The solution can be thus written as equations (3.56) and (3.57).

$$U[t] = \lambda cos(\Omega t - \alpha) + \frac{\epsilon}{32\omega^2}\lambda^3 cos(3(\Omega t - \alpha)); U \geq 0 \tag{3.56}$$

$$U[t] = \lambda cos(\Omega t - \alpha); U \leq 0 \tag{3.57}$$

The waveform obtained from the harmonic excitation is used further after filtering out the excitation frequency. The band pass filtered signal is then processed through HT and the natural frequencies for open and closed state of crack is obtained.

3.4.3 Random Excitation

For the system to be under random excitation, two parameters can be changed in the input force P_{mn}, expressed by equation (3.58)

$$P_{mn} = A_r cos\omega_r t \tag{3.58}$$

where, A_r is the random amplitude and ω_r is the random frequency. The random input vector can be generated from the software $MATLAB^{TM}$ by using 'rand' command and substituting in the set of equations. These equations can then be solved numerically as shown in section 3.5.

3.5 NUMERICAL SOLUTION

For the system to be solved numerically, the fourth order Runge-Kutta method is employed for the Duffing equation. As the solution of piece-wise equations is required
to get the response of bilinear model of breathing crack, the response is obtained in
MATLAB using bilinear equations (3.24) and (3.25). An aluminum plate ($E = 70.3\ GPa$,
$\rho = 2660\ kg/m^3$) with SSSS BC having length l, breadth b and height h equal to $0.5m$,
$1m$ and $0.01m$ respectively, is considered for numerical analysis. The initial conditions
for displacement and velocity is set to 0 and $1m/s$, respectively. The stiffness ratio α is
set to be 0.9. Figure 3.3 shows the impulse response obtained by applying Runge-Kutta
method.

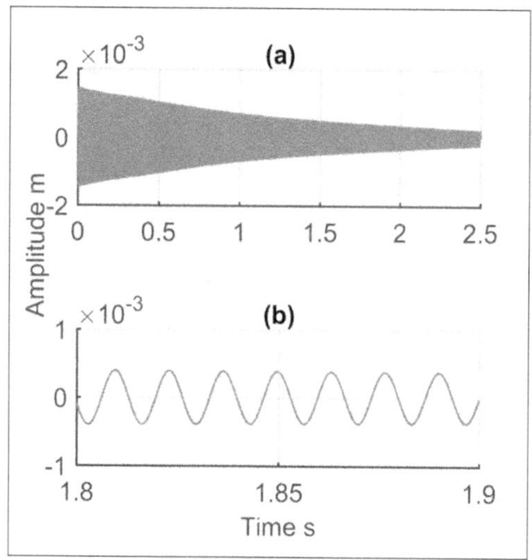

Figure 3.3: (a) Impulse response of a cracked plate (b) Zoomed area

For the harmonic input with frequency close to the natural frequency of the plate, the response wave is shown in Figure 3.4.

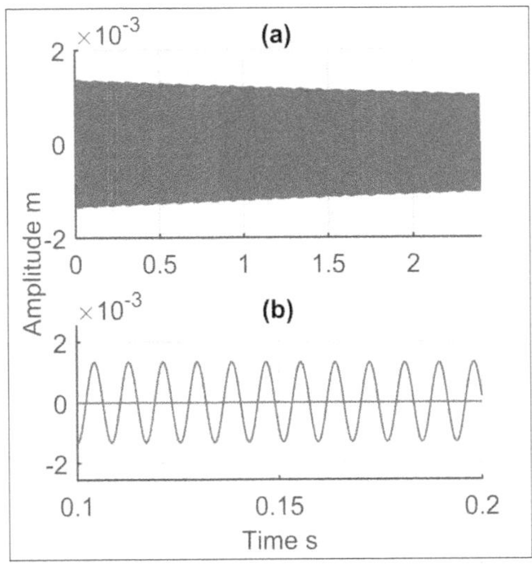

Figure 3.4: (a) Harmonic response of a cracked plate (b) Zoomed area

The response to random excitation depending upon the random amplitude and

frequency can be seen as in Figure 3.5.

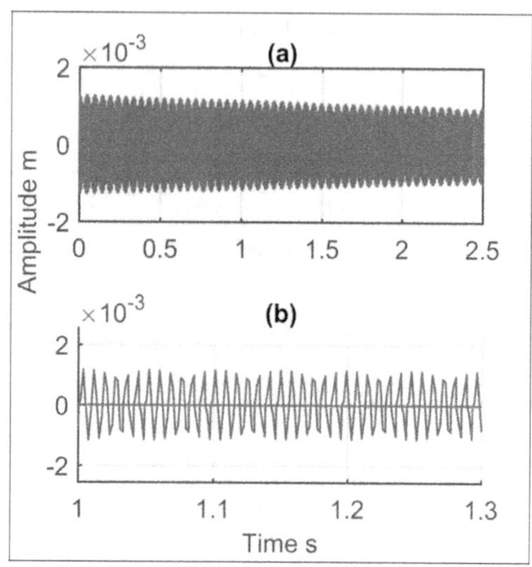

Figure 3.5: (a) Random response of a cracked plate (b) Zoomed area

The Figures 3.3-3.5 show the numerical responses to three different excitations, obtained as a result of applying Runge-Kutta method.

3.6   RESULTS AND DISCUSSION

A novel mathematical modeling has been presented in this research which uses different types of excitation to get response equations for part-through breathing crack present in a plate, in the form of piecewise equations. The bilinear dynamic characteristics of the breathing crack has been considered and the results drawn as follows:

1. Initially, the mathematical model of a plate with an all-over, part-through crack was developed as an extension to the model of an open part-through crack and the validation was done by comparing the natural frequencies obtained from the developed methodology with that of the published all-over breathing crack. The result of Table 3. 2 showed close agreement between the two.

2. The analytical results obtained directly from modal analysis for all four BCs (CCFF, CCSS, SSSS and CFFF) showed that the drop in natural frequencies of the breathing crack was lesser than that of an open crack as observed in Table 3.1. Hence, showing lesser difference directly between the intact and damaged plate.

3. The results obtained showed that as the crack severity ratio increased from 0.05 to 0.5, the difference in natural frequency increased. Also, the difference for open crack was more than that for the breathing crack as evident from Figure 3.2. It also showed the dependence of natural frequency on the BCs and that the largest frequency difference was for the CCFF case.

4. The natural frequency of the plate with a breathing crack is greater than that of the plate with an open crack due to the fact that during vibration, the crack opens and closes. When the crack is closed it behaves like an intact plate. Hence, reducing the open crack time which indirectly means that half of the time it behaves as an intact plate and half of the time as cracked as calculated by Equation 3.26.

5. The analytical solution using the method of multiple scales was derived in the form of piecewise equations (3.45 and 3.46) for impulse excitation. It has been shown that during the positive half wave of the vibration, the crack opens and induces a cubic nonlinearity and as the crack closes during the negative half cycle, the nonlinearity approaches to zero. Thus it can be concluded that the breathing crack can be modeled as a bilinear model based on its dynamic characteristics.

6. The impulse excitation can be referred to as the free response of the system as it gives the natural frequency of the system upon settling down as evident from response equation whereas the harmonic excitation contains the excitation fre-

quency (close to natural frequency) as shown by equation (3.56 and 3.57).

7. The random excitation is considered by changing both the amplitude A_r and frequency ω_r in the response equation. This can be achieved by updating the state of the vector in the duffing oscillator in $MATLAB^{TM}$. The resulting piecewise plot obtained can then be used for further signal processing.

4.CRACK IDENTIFICATION METHODOLOGY

4.1 INTRODUCTION

In the previous chapters, the analytical and numerical response of cracked plate has been developed. The response obtained is then utilized for identification by applying the proposed methodology. This chapter presents the crack identification methodology applied on the response of the cracked plate, along with severity estimation based on HT.

4.2 HILBERT TRANSFORM (HT)

HT is a very useful operator in signal transformation and processing. It converts a real function into an analytic one hence, extending its use practically in the field of signal theory. The impact of analyzing the function brings precision and accuracy to the results and this is the reason HT is credited with giving a major boost in the signal studies.

The HT was developed by David Feldman and it explores the instantaneous characteristics of the signal i.e. the instantaneous amplitude (IA) and the instantaneous frequency (IF). The HT takes a function of a real variable and produces another real function. If a given function is assumed to be $X(t)$, then its HT is represented by $H(X(t))$, hence the integral formula is shown as equation (4.1)

$$H(X(t)) = \lim_{\epsilon \to 0} \frac{1}{\pi} \int_{|s-t|} \frac{X(s)}{(t-s)} \, ds \tag{4.1}$$

However, the integral equation does not help much mathematically in understanding the transform and so the analytic signal X is defined in terms of real and imaginary parts as equation (4.2):

$$X = x + \iota H(x) \tag{4.2}$$

where, $H(x)$ is the HT of the signal x. The HT of a signal is defined as the transform in which phase angle of all components of the signal is shifted by $\pm 90^o$ [108]. The IA is the envelope of the signal while IF can be obtained from the derivative of the phase Ψ of the analytic signal as shown in the equation (4.3).

$$IF = \omega(t_i) = \frac{d}{dx} \Psi_x(t)|_{t=t_i} \tag{4.3}$$

Different researchers have worked on HT and done extensive work on developing its theory [109]. The major contribution was done by Huang by carrying out Empirical Mode Decomposition (EMD) of the signal and then performing HT [110].

The HT has been used in the past for vibration based damage detection. It can be used as system identification i.e. it can detect the nonlinearity of the system by producing a distortion in the frequency response function (FRF) which otherwise remains unchanged, if the system is linear [111, 112]. Similarly, in time domain, the reduction in slope of the envelope of the response signal corresponds to the damping of the system [113]. The HT can give information regarding the system nonlinearity in the form of system backbone and damping curve. The tilt of the curve or backbone gives the measure of hardening or softening of the stiffness/damping parameter in the system [114]. The HT is also dependent on the signal variation and composition. The signal may be narrow band or wide band and may be mono component or multi component. The HT is applied to slow varying mono component signals. The HT can only be applied to narrow band signals and for this reason the signal is first band pass filtered.

4.3 PROPOSED METHOD

The time response obtained as a result of vibration of the cracked plate is further processed to obtain useful information from the signal. First of all, the system is excited using impulse excitation and the response $x(t)$ is captured from accelerometers attached at the mid and end point of the plate. Then the FFT of the signal is performed to obtain the natural frequency of the plate. After obtaining the frequency region of the first mode,the signal is band pass filtered. The signal is further separated into positive and negative halves and then reconstructed. The HT is applied to the separated reconstructed waves and their respective IFs are obtained. The difference between the IFs of the negative and positive halves of the wave shows that the plate has a breathing crack or not. Usually a threshold value of the difference between IFs is allotted below which the plate is considered intact.The same procedure is carried out for harmonic and random excitation. Figure 4.1 shows the procedure.

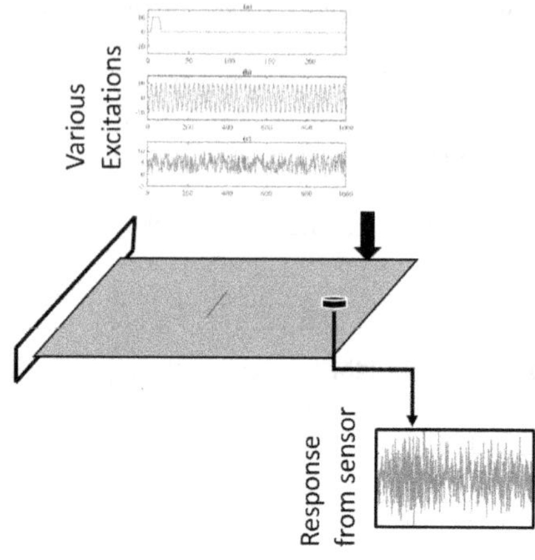

Figure 4.1: Method to obtain the response of cracked plate

The proposed procedure for carrying out the identification of crack is shown in the form of a flowchart in Figure 4.2 and the details are given in section 4.4.

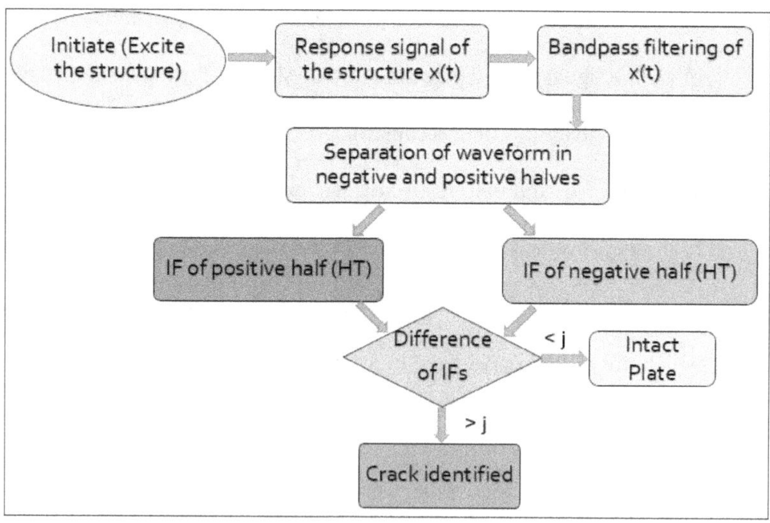

Figure 4.2: Flowchart of the crack identification methodology

The value of j represents the threshold value for the difference in IFs. The plate is considered to be intact if value of j is lesser than 0.01 except for the cantilever case in which j is lesser than 0.1. Otherwise, the plate is considered to have a crack.

4.4   STEPS OF CRACK IDENTIFICATION

The identification of the breathing crack is carried out in the following steps:

4.4.1   Response Signal from the Nonlinear Equations

The algorithm for piecewise equations for the two states of breathing crack is formulated in $MATLAB^{TM}$ and the signal is acquired. The equations may be solved numerically using the Runge-Kutta method or solved analytically by using the solution obtained in chapter 3. The resulting waveform will represent the response of a plate with the part-through breathing crack. The resulting analytical wave form for harmonic excitation is given in Figure 4.3.

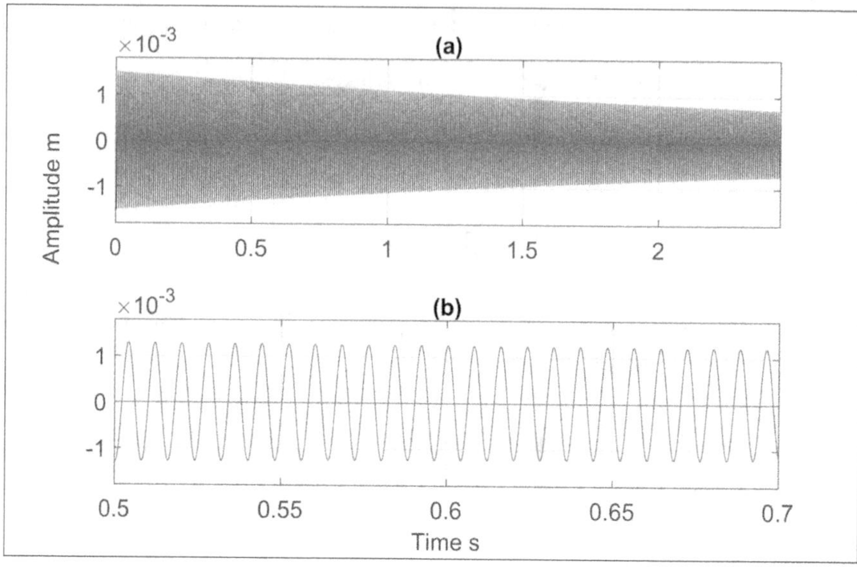

Figure 4.3: (a)Analytical response of cracked plate with harmonic excitation at 125Hz (b)Zoomed area

4.4.2 Investigation of Frequencies from Response Signal

The response from analytical equations, presents the global response of the breathing crack model. To get the local response i.e. the opening and closing of crack,the response is separated into positive and negative halves. The positive half of the wave represents the open crack and the negative half represents the closed crack. The Figure 4.4 shows the procedure of obtaining local responses.

4.4.3 Extraction of Natural Frequencies by Hilbert Transform

The HT is used to find the IF from the response of a plate with a breathing crack. In case of a beam with a breathing crack, it has been established that greater the amplitude of

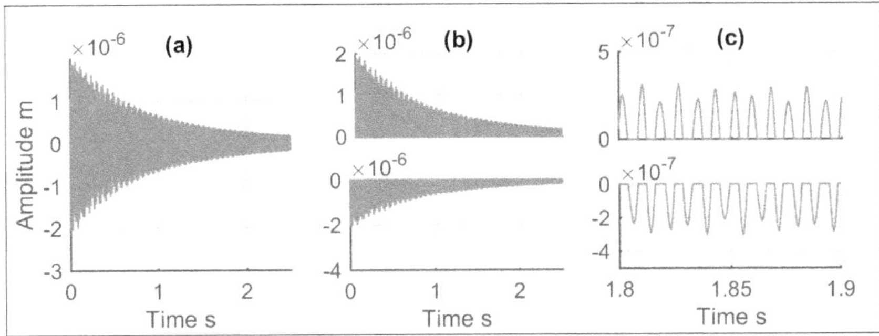

Figure 4.4: Separating the (a) Impulse response in (b) Positive and negative halves (c) Zoomed area of response in (b)

the IF, greater is the crack severity. The HT does not properly work on the half waves only, so to rectify this problem an algorithm is developed that converts the half sine waves to full sine waves for the HT to be applied accurately as shown in Figure 4.5. Firstly, the raw signal is interpolated to obtain values closer to zero. These values assist in finding the zero crossings. Then, the second half of the wave is inverted so it forms a full wave. The total time thus reduces to almost half for each region i.e. positive region and negative region. The IFs of the positive half wave and the negative half wave can be extracted by then applying the HT on both waves separately. It should be observed that the positive half has lower frequency than the negative half. Lastly, by comparing the values of both frequencies, the presence of breathing crack can be ascertained, and its severity indicated.

4.4.4 Comparison of Local Natural Frequencies

The identification method for breathing crack present in plate is developed by considering the results obtained from the system identifiers i.e. IFs. It has already been explained that the cracked plate response represents a bilinear system as its stiffness

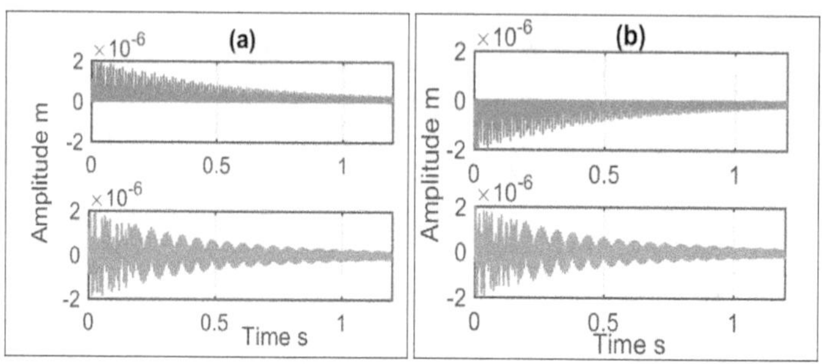

Figure 4.5: Reconstruction of (a) Positive half-wave (top), Reconstructed wave (bottom) (b) Negative half -wave (top), Reconstructed wave (bottom)

varies in two states. As the plate can behave as a bilinear oscillator under a specific mode, it is assumed that it is oscillating in its mode I. The mass and damping would remain constant. The stiffness changes whenever the plate vibration crosses the $U = 0$ point, hence changing the frequency. By observing the difference between the natural frequencies in the two regions (local), the breathing crack can be identified. If the difference is zero or nearly zero, it means that the breathing crack is not present. Whereas, the relative change in frequency will indicate presence of breathing crack. For known location, the greater the difference, the more is the severity of the crack, thus quantifying the crack as well.

It is evident from Figure 4.6, that the difference in IF's of the positive and negative

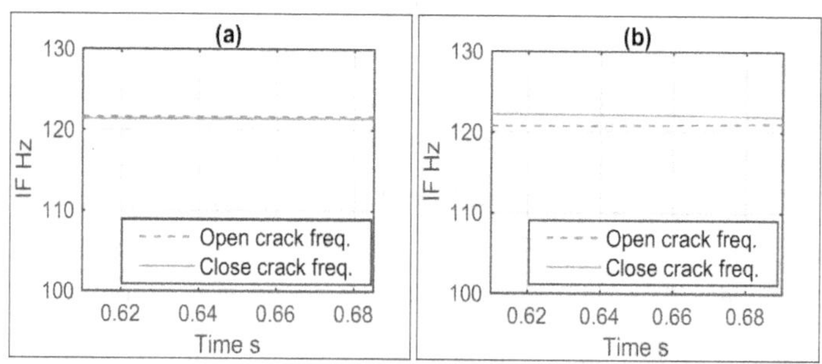

Figure 4.6: Relative change in Instantaneous Frequencies of a SSSS plate (a) Intact (b) Cracked

halves of the response signal exists in cracked case. Whereas, in intact plate the lines overlap, showing no difference. Hence, by employing the proposed method, the breathing crack can be identified successfully. Moreover, no baseline information is required for identification, as the negative half IF represents the closed state i.e. intact plate frequency.

The same procedure is applied to the responses obtained from harmonic and random excitation. However, to identify the region of first mode of vibration, FFT is performed and then the filtered signal is used for further signal processing.

4.5 BILINEAR FREQUENCIES FROM ANALYTICAL AND NUMERICAL SOLUTIONS

The Runge-Kutta method given in section 3.5 has been employed to verify the results obtained from Multiple Scale Method in section 3.4. Then the section 4.4 uses the nonlinear response of the equation 3.45 and 3.46 and after applying HT, obtains the desired frequencies. A table comparing simulated and analytical results of the bilinear frequency has been given. The method has been applied on an SSSS plate having the length l, breadth b and height h equal to 0.5 m, 1 m and 0.01 m respectively and a crack of length $2a$ in the center.

Solution Method		Crack severity ratio ($2a/l$)						
		Intact	0.05	0.1	0.2	0.3	0.4	0.5
Analyti-cal (Hz)	ω_o	123.587	123.375	122.729	121.824	121.235	120.819	120.511
	ω_c	123.587	123.587	123.587	123.5872	123.587	123.587	123.587
	ω_{br}	123.473	123.03	122.675	122.208	121.916	121.715	121.568
Numeri-cal (Hz)	ω_o	123.587	123.375	122.729	121.824	121.235	120.819	120.511
	ω_c	123.587	123.587	123.587	123.587	123.587	123.587	123.587
	ω_{br}	123.587	123.481	123.156	122.699	122.400	122.187	122.029
Simula-ted (Hz)	ω_{br}	122.87	122.869	122.866	122.853	122.84	122.819	122.800
Error % - numerical & simulated		0.580	0.495	0.236	0.125	0.359	0.516	0.631

The Table 4.1 compares the results obtained by the proposed mathematical model using both solution methodologies i.e. Multiple Scale Method and Runge-Kutta Method, to the breathing frequency obtained from the simulated model. The results show close agreement between all frequencies i.e. error less than 1%.

4.6 COMPARISON BETWEEN PERCENTAGE FREQUENCY CHANGE AND FREQUENCY RATIO

It has been established in the previous sections that the change in frequency between open and closed crack state gives us the severity of the plate. However, the results obtained from the frequency change although are a good indicator of breathing crack but the conventional method of expressing the severity with frequency is by frequency ratio. Therefore, the trend of frequency with respect to severity effect is also shown as it would be comparable with other studies. The frequency ratio is named as Instantaneous Frequency Ratio (IFR) due to the instantaneous nature of the frequency. IFR is defined as ratio of Instantaneous Frequencies of the open crack (ω_o) to closed crack (ω_c) state i.e.

$$IFR = \frac{\omega_o}{\omega_c} \tag{4.4}$$

The Figure 4.7 shows the comparison between both trends.

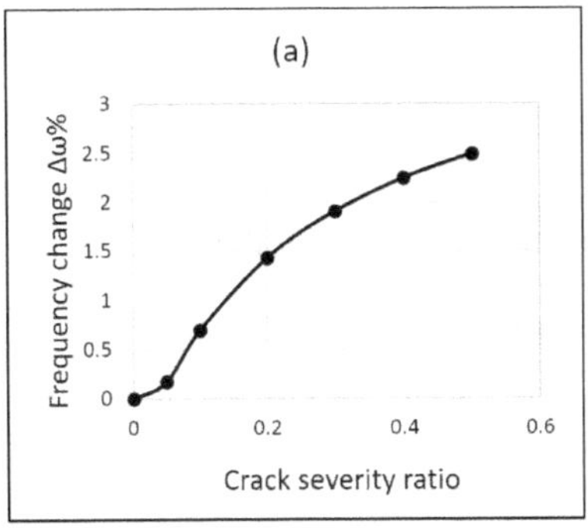
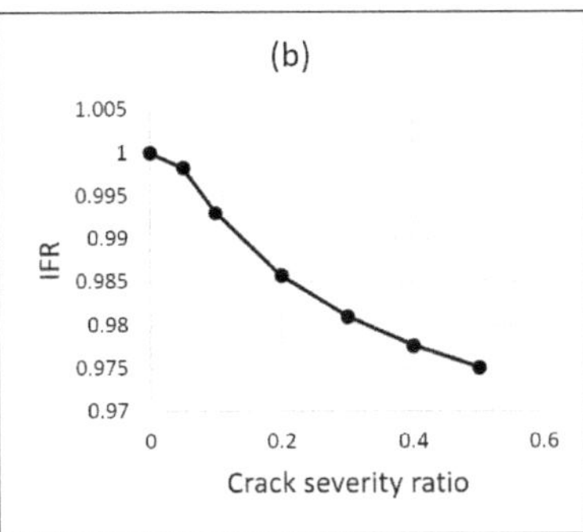

Figure 4.7: Graph trends for (a)Percentage change in frequency (b) Instantaneous frequency ratio

It can be observed that the percentage frequency change is increased on increased severity (Figure 4.7a) and the IFR is decreased on increased severity (Figure 4.7b). By comparing both graphs it can be seen that the crack is detected more efficiently in percentage frequency change which has been proposed in this research.

4.7   RESULTS AND DISCUSSION

The proposed methodology is applied on the responses obtained under different types of excitations. It has been shown that the proposed methodology uses the HT to derive the instantaneous characteristics from the time response wave of the plate having breathing crack. The results are :

1. The open crack frequency is lesser than the closed crack frequency and their difference gives the identification.

2. The crack identification can be done by taking data from a single point of sensor hence reducing the cost of extra sensors.

3. The band pass filtering allows us to find the instantaneous characteristics of the system in the desired mode.

4. During reconstruction of positive and negative halves to full waves, the total time of reconstructed signals is also reduced to half.

5. The present study focuses on first mode of vibration hence it can be said that the breathing crack identification is done in first mode which is an achievement.

6. The HT method identifies the change of frequency over time, hence giving us the information of the bilinear frequencies of the breathing crack.

7. Table 4.1 shows that the analytical values of the frequency remains same when the waveform is processed for the developed procedure, hence authenticating the proposed method.

8. The severity can be plotted either against frequency change percentage or IFR. If the $\triangle\omega\%$ is considered, its relation is directly proportional to the crack severity ratio i.e. as severity ratio increases the $\triangle\omega\%$ increases as shown in Figure 4.7(a).

9. The crack severity ratio has inverse relation with IFR i.e. as the crack severity increases, the ratio is decreased. The case for SSSS plate is shown in Figure 4.7(b).

5.FE MODEL OF CRACKED PLATE-ANSYS

The finite element models of the plate having a breathing crack with different severity ratios are modeled using the commercial software ANSYS® Mechanical APDL [115]. ANSYS provides the different types of modeling and analysis and vastly used in designing process. It can produce 3D designs and their stress analysis, deformation over time, modal analysis etc.

5.1 MODELING STAGES

The stages in modeling of the cracked plate in ANSYS are as follows:

5.1.1 Creating 3D Model of Plate

First of all, Structural is selected in the Preferences and the material properties like modulus of elasticity, density, damping ratio and poisson's ratio are defined according to the selected material in the Preprocessor option. The material is Aluminum alloy 5000 series and 20-node elements SOLID186 is used. Next the model is prepared by creating four blocks in the Volumes option.The width of the middle blocks are kept to be of crack length as shown in Figure 5.1.

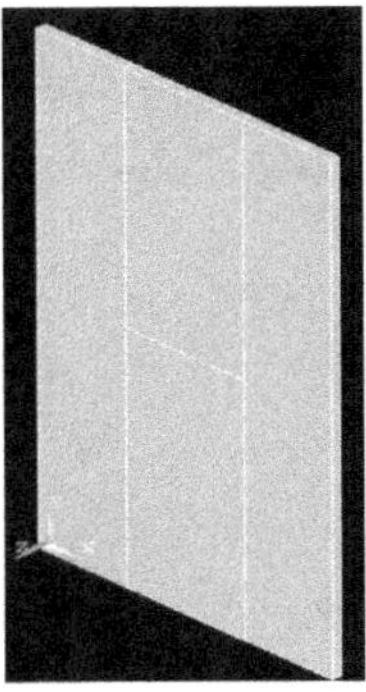

Figure 5.1: Step1: Creating volumes by 3D four blocks

5.1.2 Meshing

The blocks are then meshed.The crack is introduced after the meshing size is checked for the intact plate. The element edge length (e) is selected by doing a convergence analysis test. For each iteration, the element edge length is halved and it is shown that at $\frac{1}{8}e$, the error percentage is the lowest. Also, the simulation time is checked for different element sizes and the optimum is selected as shown in Figure 5.2.

The FE model is shown in the Figure 5.3. The middle of the plate is fine meshed as

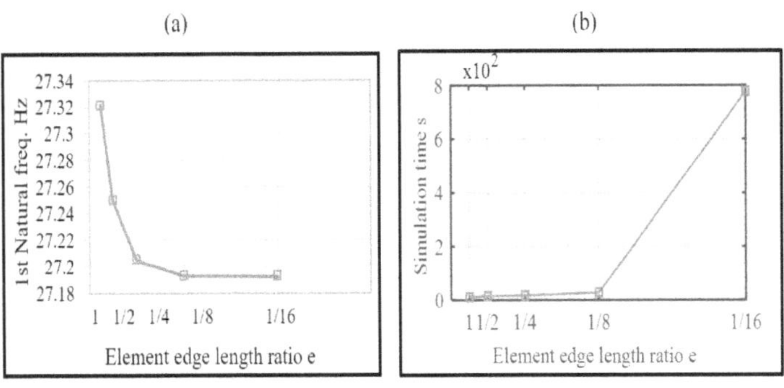

Figure 5.2: (a)Convergence test for meshing of plate (b) Optimum simulation time

crack is present in the middle. The zoomed view is also in the Figure 5.3.

Figure 5.3: Step 2: Meshed model of cracked plate with zoomed area at center

5.1.3   Creating Contact Pair

The breathing behaviour of the crack is modelled by introducing contact pair model using target and contact surfaces. This prevents the node penetration during the analysis. Contact pair is developed using TARGE170 and CONTA174 element types for 3D solid element SOLID186. The surfaces are considered frictionless and rest of the settings are set by default. The open crack state area is zoomed and presented in Figure 5.4.

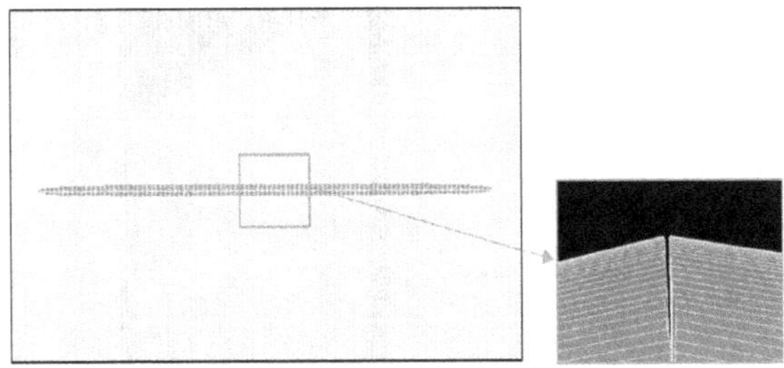

Figure 5.4: Step 3: Creation of contact pairs in crack with zoomed area showing open crack in side view

5.1.4 Applying Load and BCs

The last stage of modeling is the application of the boundary conditions. For this, the area to be constrained is selected and the displacement is set to 0 in the desired direction. For clamped-end, all the three directions (x,y,z) are set to 0 and for simply supported, only the displacement in z-direction (parallel to applied load) is constrained. The Finite Element Models (FEM) are shown in the Figures 5.5-5.8.

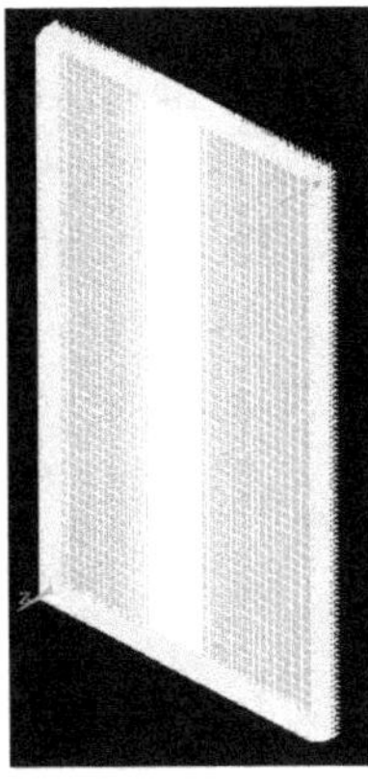

Figure 5.5: Step4(a) Application of SSSS boundaries and load

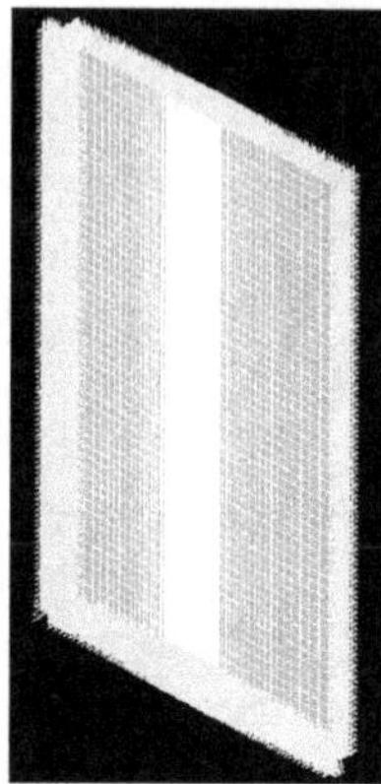

Figure 5.6: Step 4(b) Application of CCSS boundaries and load

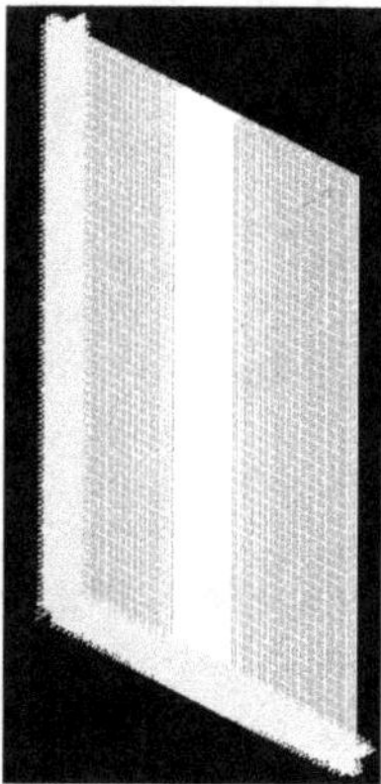

Figure 5.7: Step 4(c) Application of CCFF boundaries and load

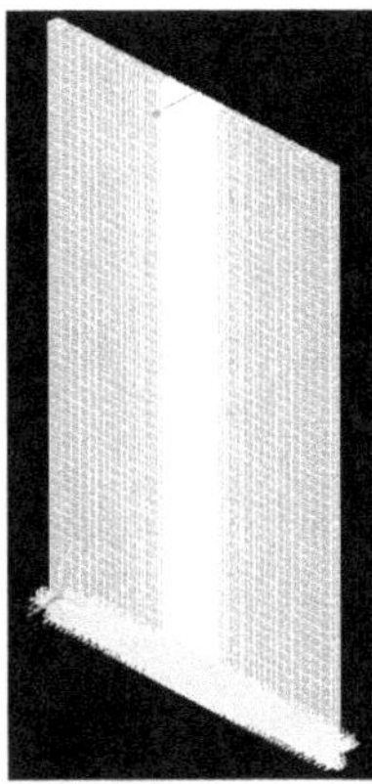

Figure 5.8: Step 4(d) Application of CFFF boundaries and load

5.2 MODAL ANALYSIS

After the creating the model, the modal analysis of the plate is performed to obtain the first five natural frequencies. The Modal analysis was performed by Block Lanczos mode extraction method.The plate is simulated for four different BCs as mentioned in Table 3.1. The Figures 5.5-5.8 show the different boundary conditions and load applied on the FEM of plate.The analysis on the plate was performed for various crack severity ratios . The extracted natural frequencies are shown in Table 5.1.

Results show that different modes show different behavior for frequency change in small cracks. It also shows that it is very difficult to differentiate between intact and cracked plate as very little frequency difference exists. So, the crack may be missed by conventional methods used for open cracks.

The first three mode shapes for different boundary conditions is given in Figures 5.9-5.12. The red colour indicates the highest deformation and blue, the minimum deformation. It is observed that when the crack comes in red region, as in SSSS case as

Table 5.1: Modal frequencies from simulated model

BCs	Mode	Simulated Frequencies (Hz) of crack ($2a/l$) model						
		0	0.05	0.1	0.2	0.3	0.4	0.5
CCFF	I	42.5024	42.5023	42.5022	42.5015	42.5010	42.4998	42.4990
	II	89.6807	89.680	89.6795	89.6742	89.670	89.6616	89.6560
	III	181.410	181.409	181.408	181.404	181.400	181.396	181.390
CCSS	I	177.433	177.430	177.429	177.413	177.400	177.374	177.36
	II	251.063	251.061	251.062	251.058	251.050	251.047	251.04
	III	378.016	378.009	377.988	377.874	377.770	377.597	377.47
SSSS	I	122.870	122.869	122.866	122.853	122.840	122.819	122.80
	II	196.040	196.040	196.040	196.040	196.040	196.040	196.040
	III	318.576	318.571	318.550	318.447	318.352	318.193	318.070
CFFF	I	8.5626	8.5625	8.5624	8.5619	8.5615	8.5608	8.5602
	II	53.269	53.268	53.266	53.253	53.242	53.223	53.209
	III	149.38	149.38	149.38	149.38	149.38	149.38	149.38

shown in Figure 5.9, it opens to its maximum. Whereas, when the nodal line appears

on the crack location, no change in frequency is observed as can be seen from Table

5.1, SSSS case mode II.

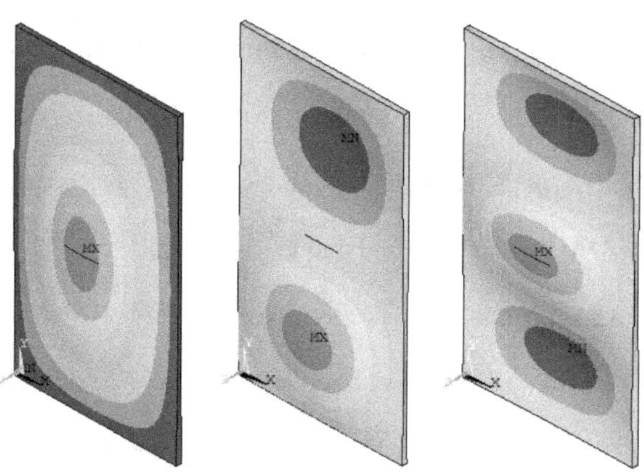

Figure 5.9: First three modes of SSSS

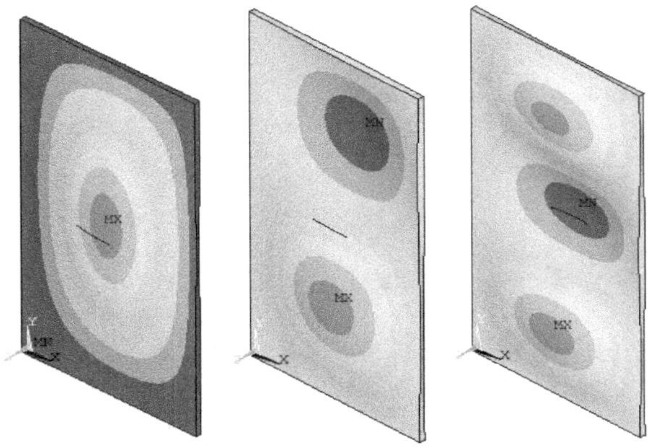

Figure 5.10: First three modes of CCSS

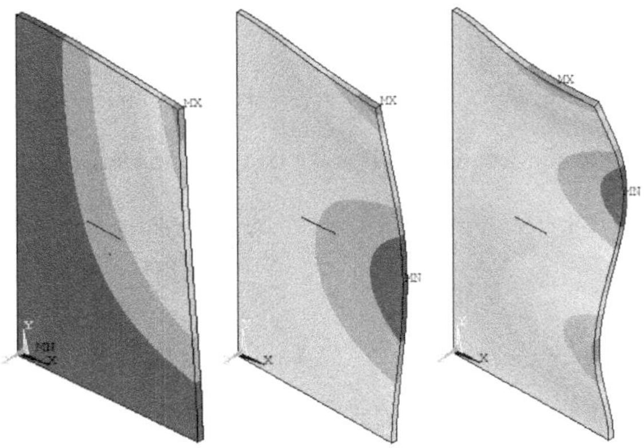

Figure 5.11: First three modes of CCFF

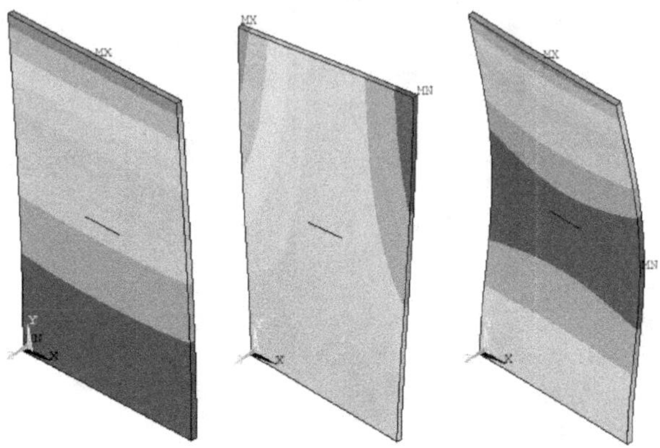

Figure 5.12: First three modes of CFFF

It should be noticed that there is no change in frequency at different crack severities in mode II of SSSS and mode III of CCFF. This is due to the presence of nodal line at the crack location in the mode II and mode III, as evident from Figure 5.13. Moreover, the open crack state during hogging can be observed in mode III as shown in Figure 5.14.

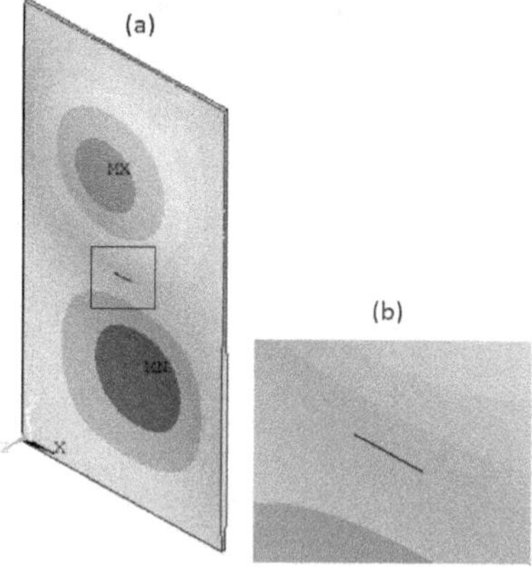

Figure 5.13: (a) Closed crack at the nodal line in Mode II of SSSS plate (b) Zoomed area at nodal line

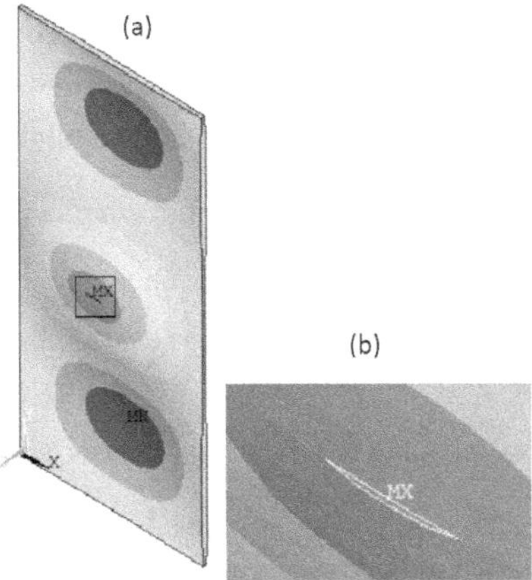

Figure 5.14: (a) Open crack in SSSS plate in Mode III (b) Zoomed area at crack

By observing the modes one can find out that during which mode the crack opens and closes and during which mode, it remains closed and the plate behaves as intact plate. Hence, modal analysis before performing transient analysis proves to be very useful.

5.3   TRANSIENT ANALYSIS UNDER DIFFERENT EXCITATIONS

Transient analysis is done when the change in amplitude of a signal with respect to time is needed. As the breathing phenomenon is a process in which the frequency changes with time, therefore instead of carrying out the usual harmonic analysis, transient analysis is performed.

The crucial step in doing the transient analysis is the selection of time step, $\triangle t$. The time step $(\triangle t)$ is calculated by multiplying the range of the first five natural frequencies with 2.56 and then taking its inverse. In this case the range is taken to be $800 Hz$ and so the time step is calculated to be 0.4882 ms.

Solution is carried out by the Newmark scheme for small deflection option and response is obtained through the Time-History Postprocessor. The transient tests are run for three types of excitations as shown in Figure 5.15. The value of X which is 49.98 in Figure 5.15(e) is nearly equal to 50Hz showing the harmonic excitation frequency.

- Impulse excitation: Load of 10N is applied for a very short duration of time and then removed.

- Harmonic excitation: A sine input of 50Hz is applied for 2.5s

- Random excitation: Loads of different amplitudes and frequencies are applied for different time intervals on the plate.

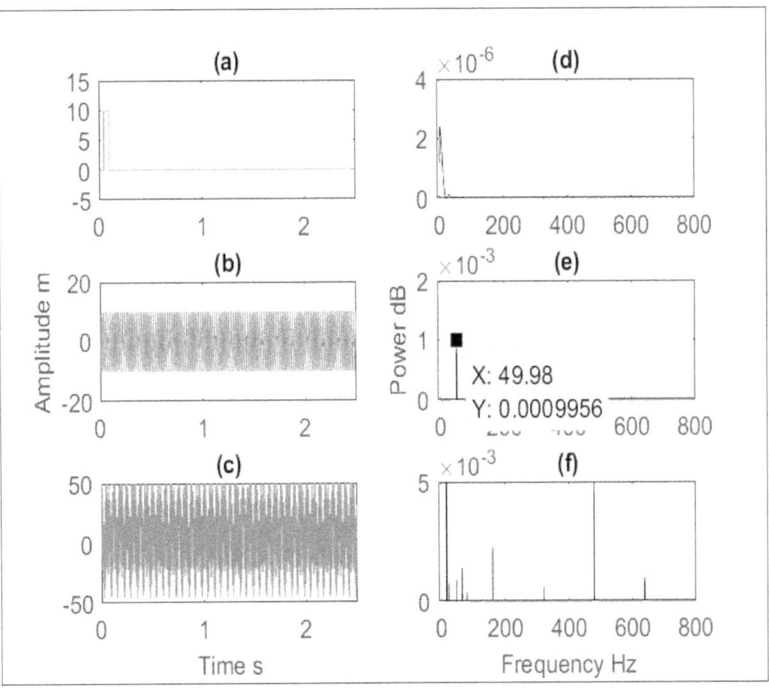

Figure 5.15: Different loading conditions(a)Impulse(b)Harmonic(c)Random time waveform and (d) Impulse FFT (e) Harmonic FFT (f) Random FFT

5.3.1 Impulse Excitation

The impulse excitation is applied to an aluminum plate having specifications as shown in Table 5.2.

Table 5.2: Specifications of plate under Impulse Excitation

Parameters	Values
Length, l	$0.5\,m$
Breadth, b	$1\,m$
Thickness, h	$0.01\ m$
Poison's ratio, v	0.33
Modulus of elasticity, E	$70.3\ GPa$
Density, ρ	$2660\ kg/m^3$
Damping Ratio, μ	0.08
Load, p	$10\ N$

The frequencies obtained from simulated time responses after applying the proposed HT are given for impulse loading in Table 5.3. These simulated experiments on plate have been performed under four different BCs considering six different crack severity ratios. The record length of the signal was 2.5 s. A load of 10 N was applied at $(x_0, y_0) = (0.375, 0.75)$ for CCFF, CCSS and SSSS, $(x_0, y_0) = (0.25, 1)$ for CFFF. The figures showing the impulse responses for various boundary conditions for intact and cracked plate (0.5 crack severity ratio) are shown in Figures 5.15-5.18.

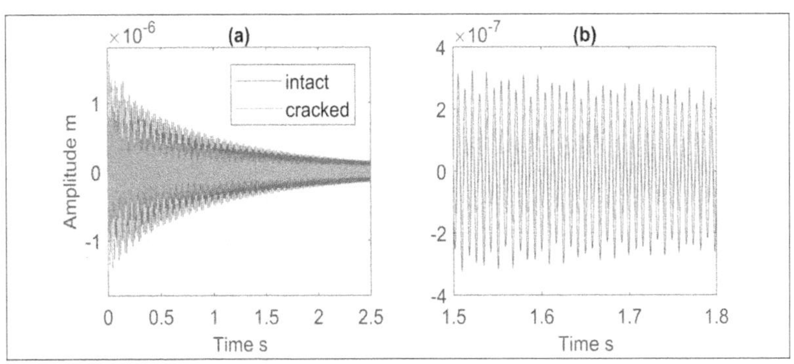

Figure 5.16: (a) Impulse response of SSSS plate (b) Zoomed area

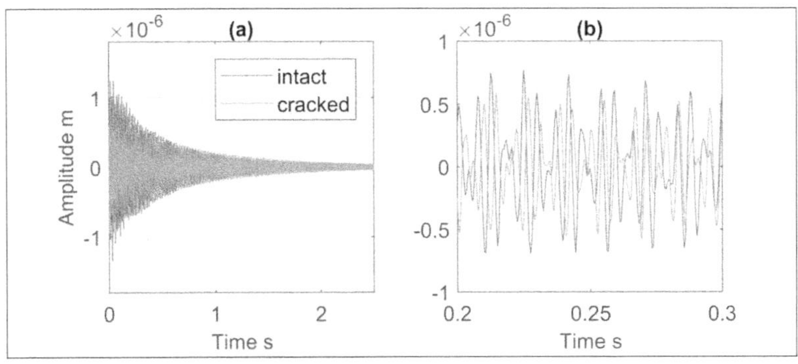

Figure 5.17: (a) Impulse response of CCSS plate (b) Zoomed area

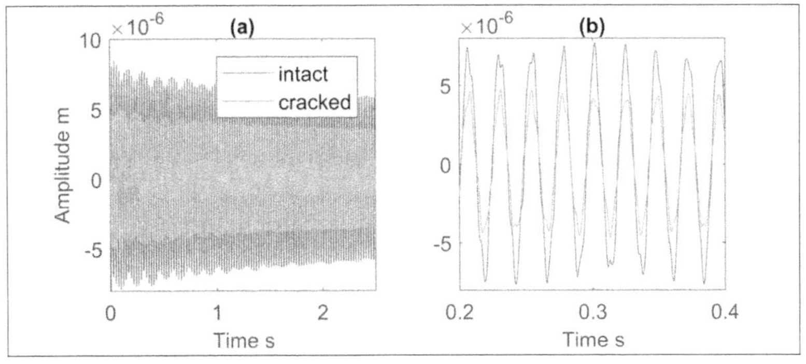

Figure 5.18: (a) Impulse response of CCFF plate (b) Zoomed area

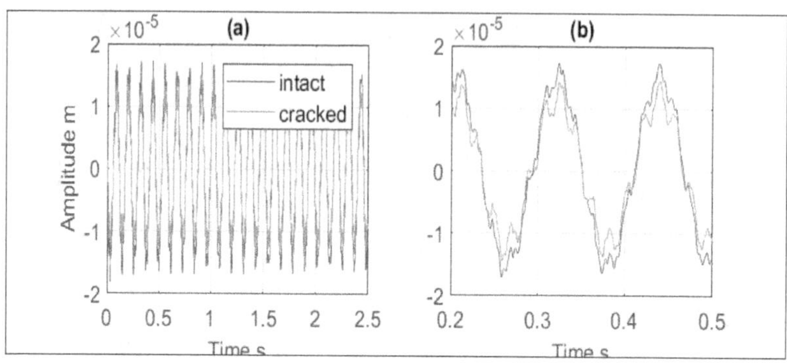

Figure 5.19: (a) Impulse response of CFFF plate (b) Zoomed area

The impulse responses show that the amplitude of the time responses are reduced when the plate has a crack. Moreover, the amplitude displacement of the cantilever plate is much more then the other BCs.

Table 5.3: Frequencies (Hz) obtained using HT

BCs		Crack severity ratio ($2a/l$)						
		0	0.05	0.1	0.2	0.3	0.4	0.5
CCFF	ω_0	42.4651	42.4650	42.4577	42.4328	42.4268	42.4104	42.4088
	ω_c	42.4657	42.525	42.4677	42.677	42.678	42.6747	42.678
	$\left|\frac{\omega_c - \omega_o}{\omega_c}\right|$ %	0.00141	0.00141	0.02354	0.5722	0.5881	0.6193	0.6307
CCSS	ω_0	173.309	173.216	173.167	173.081	172.987	172.748	172.62
	ω_c	173.299	173.443	173.444	173.441	173.768	173.768	173.781
	$\left|\frac{\omega_c - \omega_o}{\omega_c}\right|$ %	0.0057	0.13087	0.1599	0.2077	0.2623	0.5869	0.6683
SSSS	ω_0	121.525	121.500	121.311	121.213	121.187	121.032	120.963
	ω_c	121.500	121.631	121.670	121.786	121.771	121.670	121.7625
	$\left|\frac{\omega_c - \omega_o}{\omega_c}\right|$ %	0.0256	0.1078	0.3586	0.4707	0.4796	0.5244	0.6603
CFFF	ω_0	8.5425	8.5265	8.5165	8.4023	8.3856	8.3618	8.3054
	ω_c	8.5959	8.6042	8.6242	8.7102	8.7335	8.7727	8.7914
	$\left|\frac{\omega_c - \omega_o}{\omega_c}\right|$ %	0.62122	0.9030	1.2488	3.5349	3.9835	4.6838	5.5281

Table 5.3 shows that as the severity of the crack increases, the change in positive and negative half increases. This change is more than the breathing crack frequency change directly obtained from modal analysis. Also, by this method prior knowledge of plate's natural frequency is eliminated as it is based on the difference between two values from the same response obtained at that time.

The proposed method has been effectively applied on the transient responses consid-

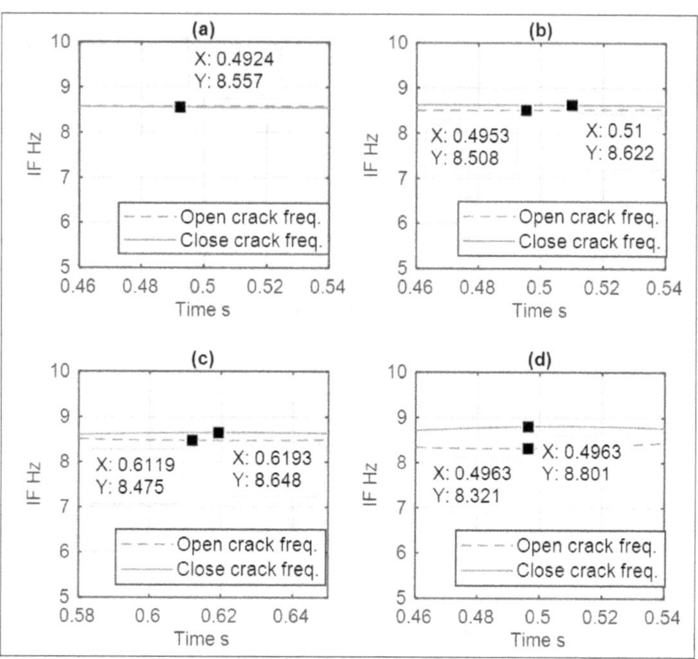

Figure 5.20: Relative change in IF's in a CFFF plate under impulse load with crack severity ratios (a) Intact (b) 0.3 (c) 0.4 (d) 0.5

ering four BCs and various crack severity ratios. It has been shown that in contrast to an intact plate, a breathing crack shows difference in IFs in the positive and negative half as shown in Figure 4.6(a) and (b). It is also indicated that the relative change in frequencies between the negative and positive halves of the cracked plate's response signal varies with crack severity ratio. The difference in IFs between two halves of a response wave is directly proportional to the increase in length of the crack as shown in Figure 5.19(a) to (d). The dashed line shows the IF obtained at the positive half while the solid line indicates the IF at the negative half. It can be observed that the closed crack frequency is a bit greater than the open crack frequency. Also, the difference between the two IFs increases as the crack severity increases.

The method is further applied to three more BCs and six severity cases to check for robustness. The IFs were obtained for all cases and then compared with the direct method. It can be observed from Figure 5.20(a) to (d) that the relative difference in natural frequencies for various crack severities is enhanced several times as compared with direct method. Also, it can be seen that the increase in trend for various BCs is different indicating the dependence of the breathing of crack on the nature of boundary of the system. It can be expressed that the proposed method is a better indicator of breathing crack and severity measure than the conventional direct frequency change method.

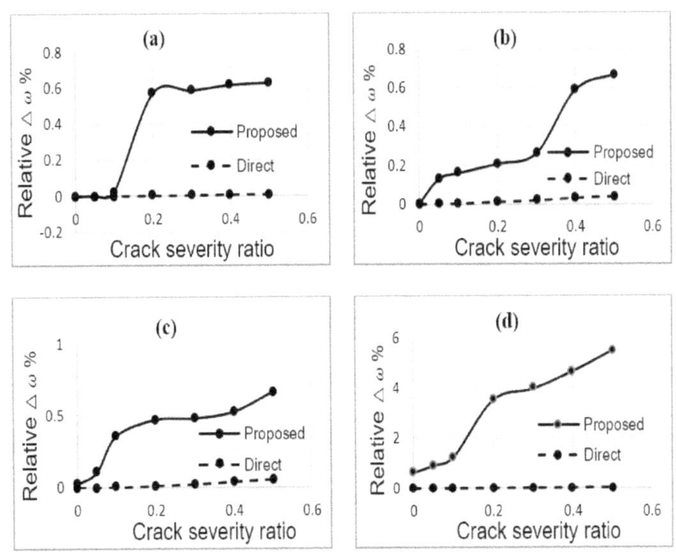

Figure 5.21: Comparison between Proposed and Direct method on increasing crack severity of plate with (a) CCFF (b) CCSS (c) SSSS (d) CFFF BCs

5.3.2 Harmonic Excitation

The same procedure is employed for harmonic excitation in case of cantilever plate with length 486 *mm*, width 250 *mm* and thickness 7.5 *mm*. The material properties are the same as that in Table 5.2.The natural frequencies for the considered plate for intact case is shown in Table 5.4.

Table 5.4: Modal frequencies for intact plate

	First seven natural frequencies (Hz)						
Intact Plate	1st	2nd	3rd	4th	5th	6th	7th
	27.2048	111.641	169.095	365.19	473.303	695.717	706.848

The input and output waveform of the harmonic excitation with their respective FFTs are shown in Figure 5.21. The values indicated in Figure 5.21(b) indicates the excitation frequency is nearly 50Hz and the value indicated in Figure 5.21(d) indicates the first natural frequency i.e. 27.2, obtained by taking FFT of the response signal which is the same as obtained by modal analysis (Table 5.4). A comparison of plots of the harmonic responses for the case of intact cantilever plate with 0.4 crack severity ratio is shown in the Figure 5.22. It can be seen that the amplitude for the cracked case is lesser than that of the intact plate.

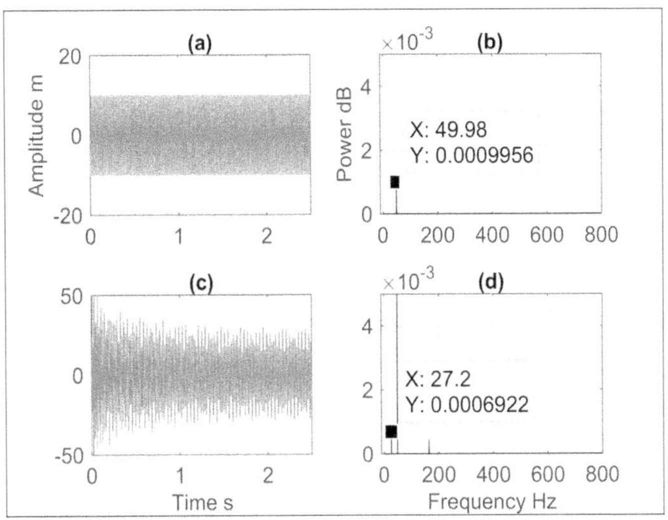

Figure 5.22: (a) Harmonic excitation (b) FFT of input (c) Harmonic response (d) FFT of output

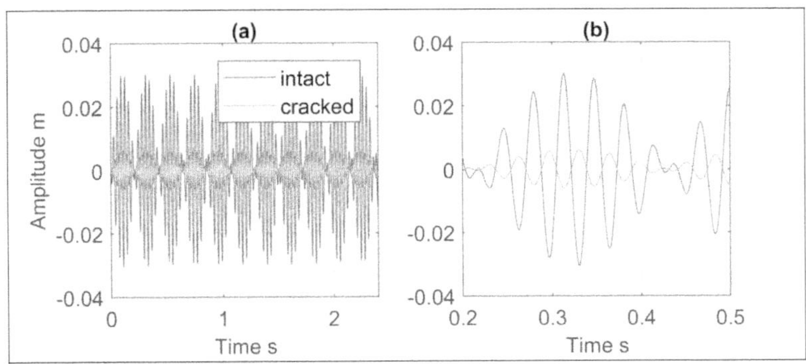

Figure 5.23: (a) Harmonic response of CFFF plate (b) Zoomed area

The crack is then introduced with crack severity ratio as large as 0.5 (50%) and as small as 0.08 (8%). The Figure 5.23 shows the IFs obtained from the plate under harmonic excitation.

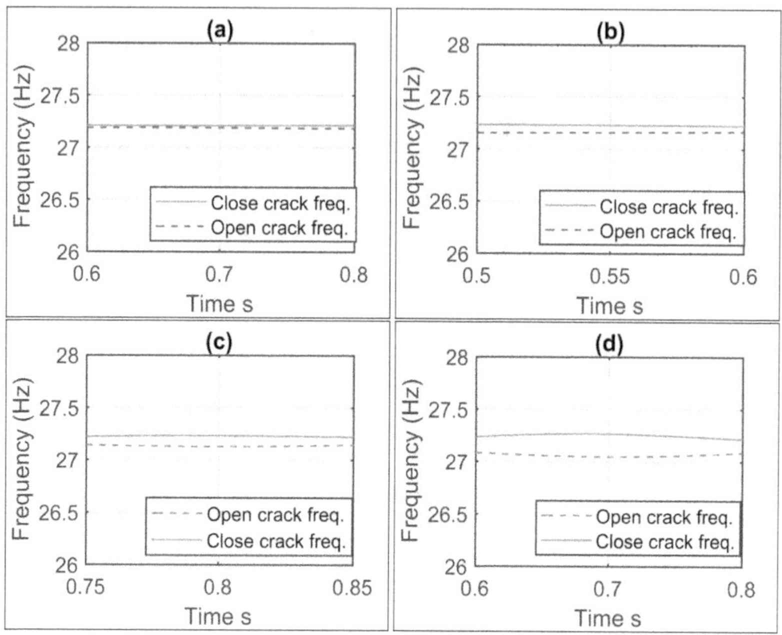

Figure 5.24: Relative change in IFs in a CFFF plate under harmonic excitation with crack severity ratios (a) Intact (b) 0.1 (c) 0.2 (d) 0.4

It can be deduced from the Figure 5.24 that even under harmonic excitation, the IFs give the same increase in difference as was for impulse excitation. Hence, proving the robustness of the proposed method. The relative frequency change can be found and its trend is shown in the Figure 5.24.

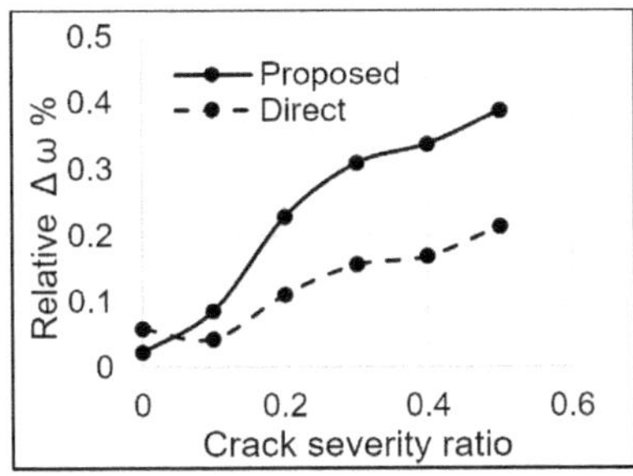

Figure 5.25: Comparison between Proposed and Direct method for CFFF plate under harmonic excitation

5.3.3 Random Excitation

The random loading is applied on the plate and the responses obtained. The simulated random response for cracked cantilever plate with 0.4 severity ratio is shown in the Figure 5.25.

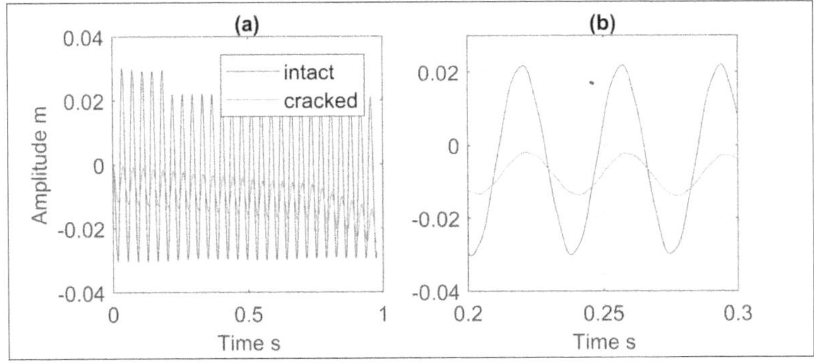

Figure 5.26: (a) Random response of CFFF plate (b) Zoomed area

If the random white Gaussian noise is used as input, the responses can be observed as shown in Figure 5.26.

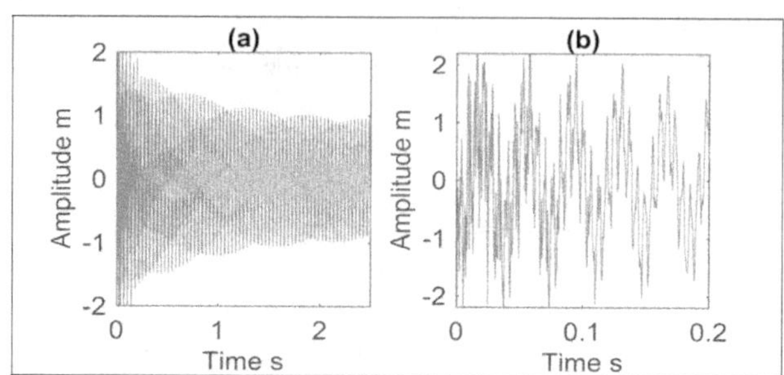

Figure 5.27: (a) Random response of CFFF plate (b) Zoomed area

The IFs are obtained according to the proposed methodology and can be shown as
in the Figure 5.27.

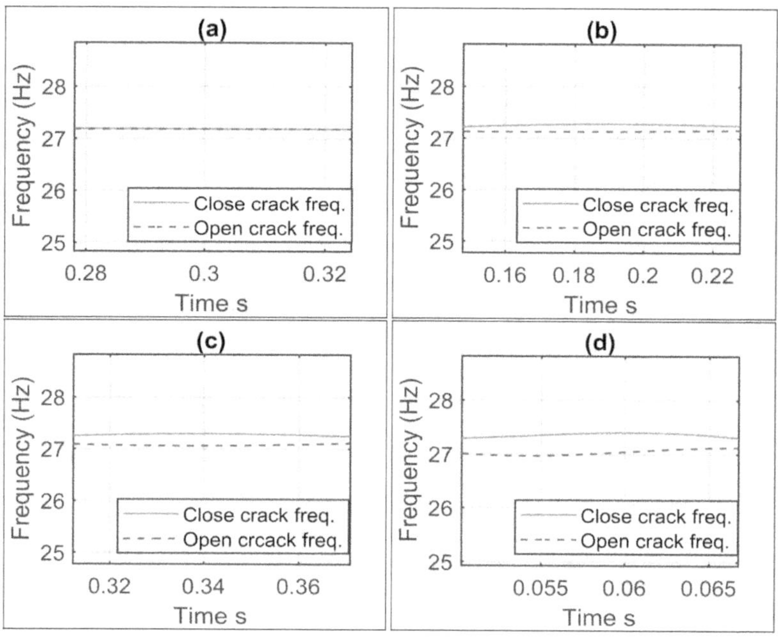

Figure 5.28: Relative change in IFs in a CFFF plate under random load with crack severity ratios (a) Intact (b) 0.1 (c) 0.2 (d) 0.4

The random responses are unpredictable so its relative change in frequency is also not always increasing as in the case of impulse and harmonic. The percentage change with respect to severity is shown in the Figure 5.28.

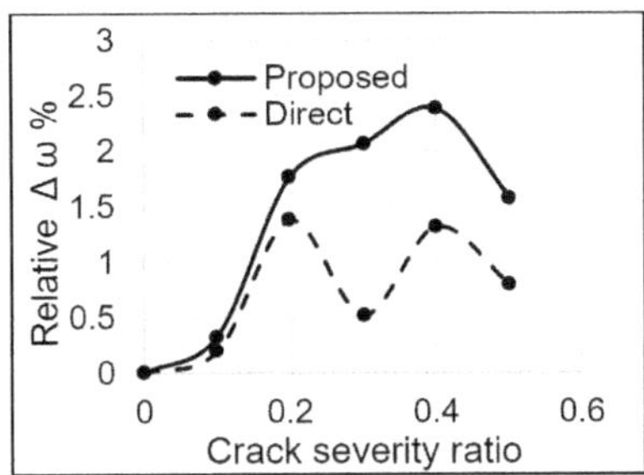

Figure 5.29: Comparison between Proposed and Direct method for cantilever plate under random excitation

It can be seen from Figure 5.28 that the crack is identified successfully for all severity cases but the trend is not smooth.

5.4 RESULTS AND DISCUSSION

A comprehensive study concerning the simulations on an aluminum plate with different BCs has been performed. The transient analysis for different excitations has also been carried out in the section 5.3. The proposed method has been applied and the results established for various crack severity ratios. The graphs have been plotted that show that with increase in severity ratio, the relative change percentage in frequency increases. For a better understanding of the response of plate under different excitations, a comparison of the proposed method among different excitations for the cantilever case has been shown in Figure 5.29.

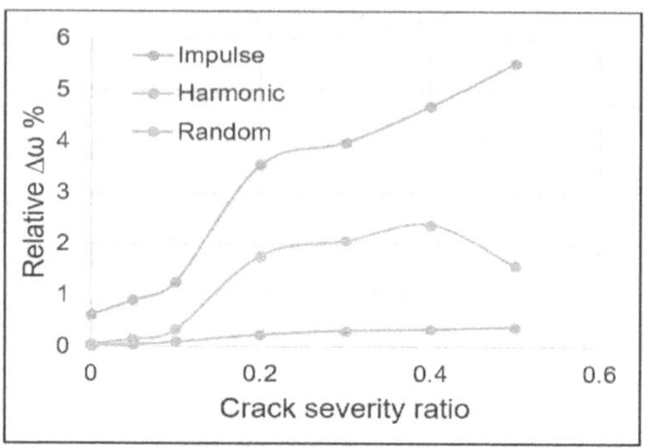

Figure 5.30: Comparison between results of different excitations

It is evident from Figure 5.29 that the proposed method works under all three excitations. However, the crack identification is best done with impulse excitation. On observing the graph closely for the three plots, it can be established that harmonic excitation predicts the best severity ratio as its graph is linearly increasing whereas in impulse and random the increase is not constant.

The other conventional method to access the crack severity is the frequency ratio. In this study, the frequencies are instantaneous, so IFR is used. The graphs of IFR for the proposed and direct methods are obtained and it is observed that both the ratios have similar trends, as shown in Figure 5.30. The dotted lines show the drop in IFR for the direct method i.e. IFR between the plate with breathing crack and the intact plate (ω_{br}/ω_c). The solid lines show the proposed method i.e. IFR between the open crack and closed crack in the plate, (ω_o/ω_c). Here, ω_{br} is the breathing frequency, ω_c is the closed crack frequency and ω_o is the open crack frequency. It can be seen that the drop in IFR is much greater for the proposed method than the direct method. Hence, giving a better indication for damage.

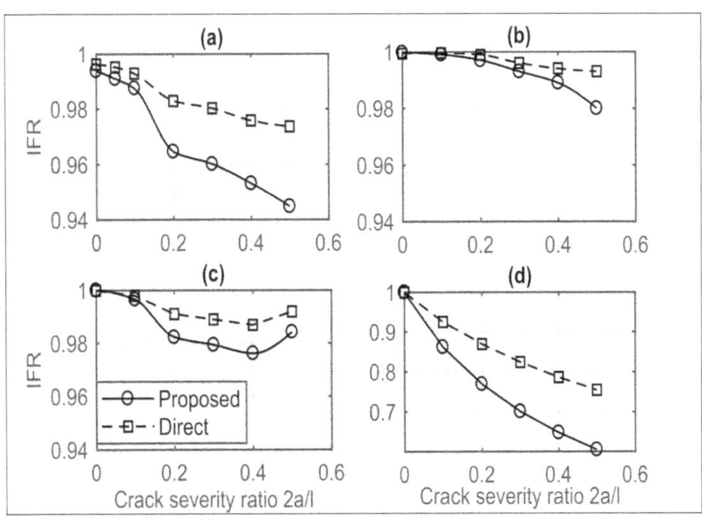

Figure 5.31: Comparison of IFR from simulated model under (a) Impulse (b) Harmonic (c) Random (d) Mathematical model

It is observed that the trend of IFR drop is similar for impulse and harmonic loading i.e. constant decrease in ratio with increase in crack severity. Also, the impulse loading shows greater drop in IFR as compared to harmonic loading. This indicates that the methodology works best in case of impulse loading and good for harmonic excitation. To check the accuracy of the method, a random size crack was modelled and the proposed methodology was applied on it. The crack size was measured from the graph by locating the IFR point on the slope. The Figure 5.30 shows that the point lies exactly on the same point as the size modelled. Hence proving the effectiveness of the proposed method.

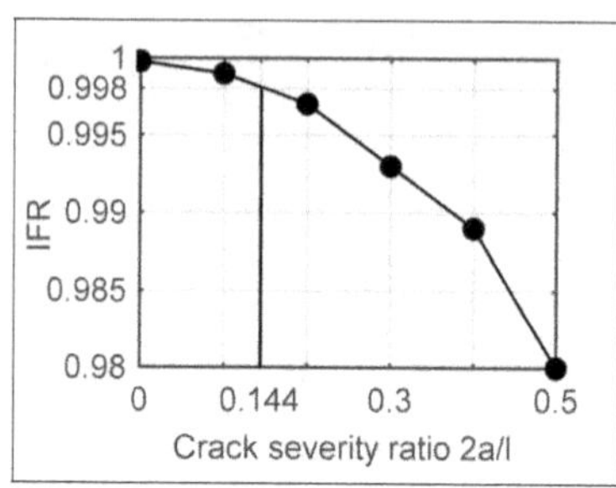

Figure 5.32: Determination of crack size from IFR

The crack size was 36mm which is 0.144 crack severity ratio. The excitation was harmonic as it is proved to be best for severity estimation whereas impulse is best for identification.

6.EXPERIMENTATION

The proposed methodology has been validated by the experimental work. The experimental analysis comprises of six steps, which are as follows:

i Preparation of the test specimen.

ii Modal analysis to find the resonant frequencies of the structure.

iii Exciting the structure under various loads such as impulse, harmonic and random.

iv Analyzing the response from various excitation.

v Applying the HT.

vi Severity estimation of the crack.

Figure 6.1 shows the complete experimental process for damage identification and severity estimation.

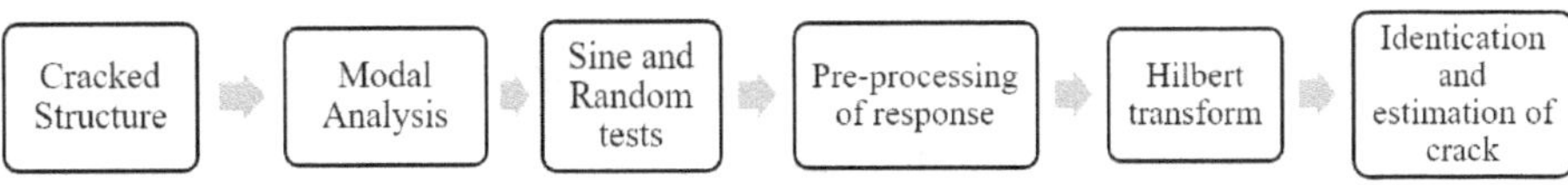

Figure 6.1: Step-wise crack identification and severity estimation process in a structure

6.1 DEVELOPMENT OF BREATHING CRACK

The test plate with dimensions and properties, given in Table 6.1, is mounted on a vertical milling machine to introduce a central part-through crack in the plate. To do this, a saw blade of thickness and diameter as 200 μm and 60 mm, respectively is mounted in the arbor of the milling machine with an attachment. This makes a 220 μm wide crack. The cutting speed is set to be 80 rpm and the feed is set at 5 mm/min. The reference surface of the plate is set as the starting position. The length of each step is 0.7 mm. The depth of the crack is achieved in 5 steps (0.7$mm \times$ 5 = 3.5 mm). The length of the crack to the plate width ratio is first set to 0.25, then increased to 0.40, and then to 0.50 as shown in Figure 6.2.

Table 6.1: Specifications of plate

Length l (mm)	486
Width b (mm)	250
Thickness h (mm)	7.5
Depth of crack d (mm)	3.5
Young's Modulus E(GPa)	70.3
Poisson's ratio v	0.33
Density ρ kg/m^3	2660
Damping ratio μ	0.08

6.2   EXPERIMENTAL SETUP

The shaker machine is used to vibrate the plate at different frequencies. The test specimen is mounted on the armature of the Electrodynamic shaker (Model no. M544A/GT1200M on its fixed end as shown in Figure 6.3(a). Three types of tests are run on the plate (sine, shock, and random). Two accelerometers (Senz 2225 100g sensitivity) are used to acquire data near the free end of the plate and another on the armature of the shaker. The electrodynamic shaker is attached to the data acquisition (DAQ) system having Vibration View software, as shown in Figure 6.3(b).

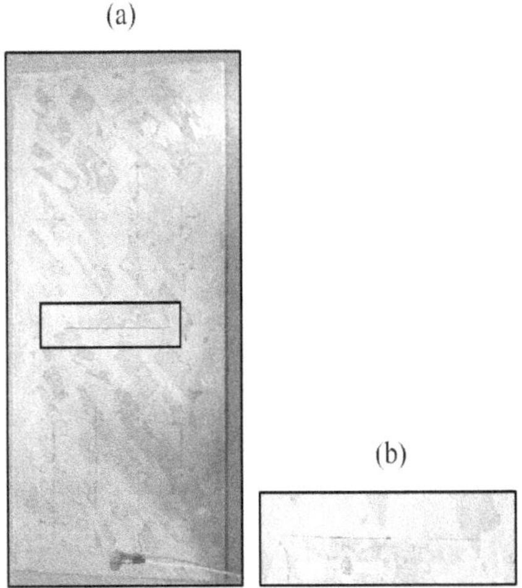

Figure 6.2: (a)Plate with 0.5 crack severity ratio (b)Zoomed area at crack

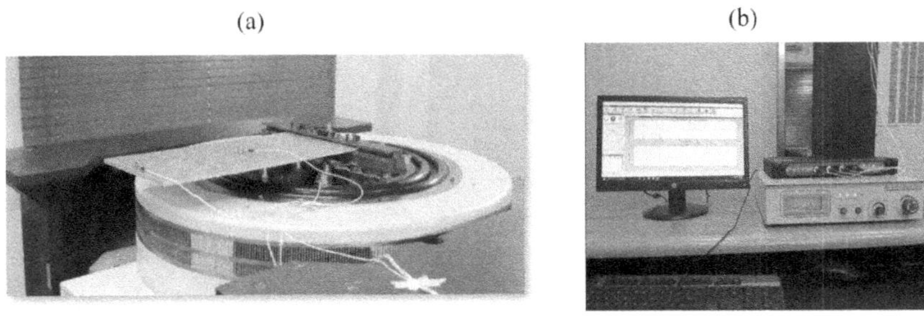

Figure 6.3: Lab testing equipment (a) Shaker machine with cantilever plate (b) DAC

6.3 DIFFERENT EXCITATIONS

In this thesis, three main types of excitations are used for investigation of the proposed methodology i.e. impulse, harmonic and random.

6.3.1 Impulse Excitation

For transient live test, the load is applied for a short duration and the response obtained. Figure 6.4 (a) shows the input and 6.4 (b) the output waveform. Their respective fourier transforms are shown in (c) and (d). The test is run for intact, 0.25, 0.4, and 0.5 crack severity ratios.

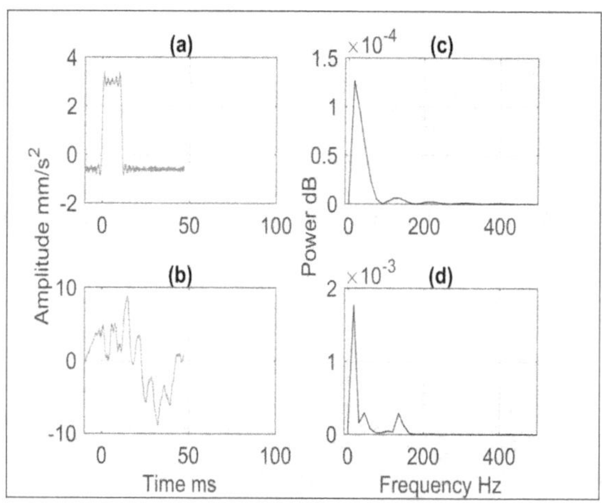

Figure 6.4: (a) Impulse excitation (b) Response wave (c) FFT of input (d) FFT of response

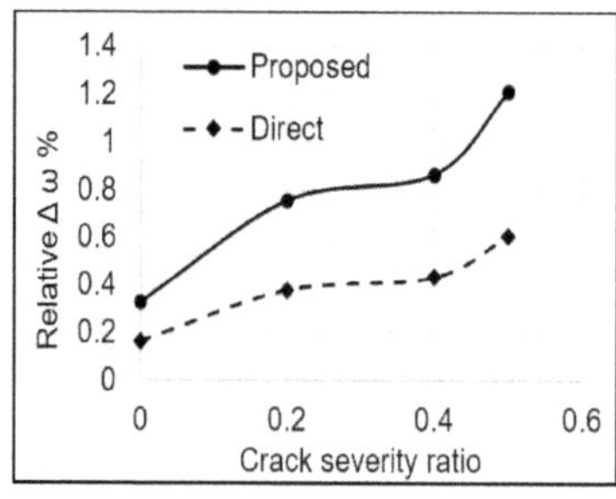

Figure 6.5: Relative frequency change percent under impulse excitation

Figure 6.5 shows the relative frequency change percent of the CFFF plate under impulse loading for the direct method and the proposed method. It can be seen that the trend of the graph for experimental (Figure 6.5) and the simulated models (Figure 5.20(d)) agree well with each other. Although experimentally, the change is greater than for simulated models. This maybe due to the effect of external factors like gravity on the plate.

6.3.2 Harmonic Excitation

For harmonic loading, two types of tests were performed. One was the sine sweep with range 5-800 Hz and sweep rate of 100 Hz/min and the other was the transient live test in which the plate was excited at 50 Hz. The sampling frequency was set to 2048. The input waveform from the harmonic loading is shown in Figure 6.6(a) and its corresponding response is shown in Figure 6.6(b), respectively.

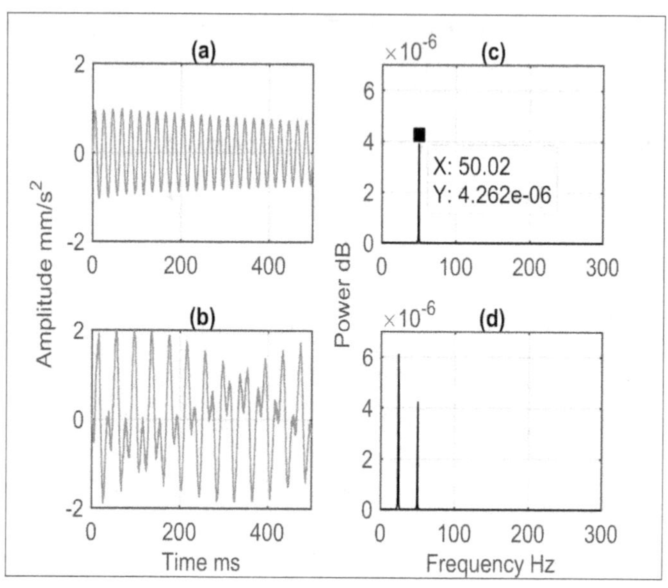

Figure 6.6: (a) Harmonic excitation (b) Response wave (c) FFT of harmonic input (d) FFT of response

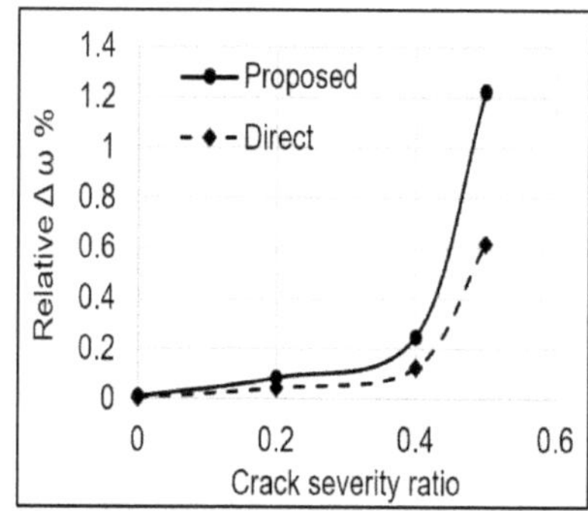

Figure 6.7: Relative frequency percent change under harmonic excitation

The FFT of the temporal signals are taken and their comparison can be seen in Figure 6.8. It is observed that due to crack, sidebands are formed. The greater the severity, the larger are the sidebands.

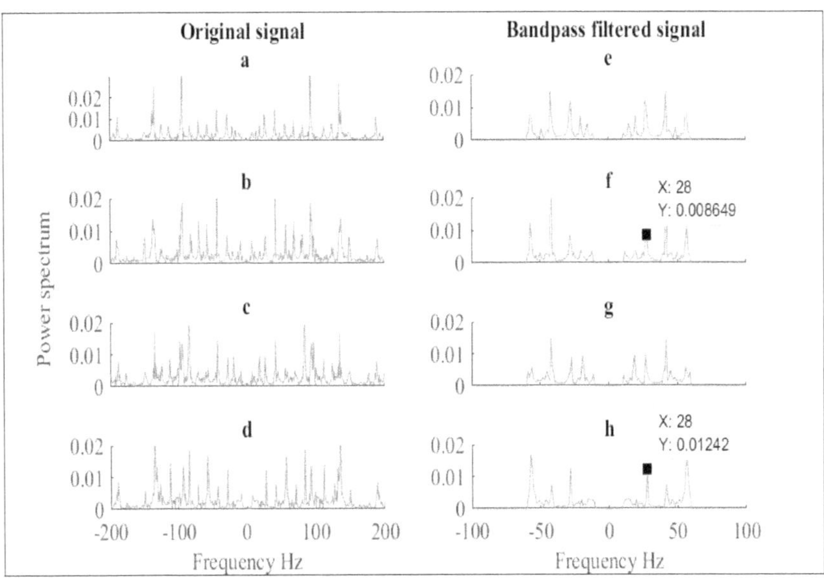

Figure 6.8: Comparison of power spectrums of (a, e) Intact plate (b, f) 0.25 (c, g) 0.40 (d, h) 0.50 crack severity ratios

Although the occurrence of sidebands may point towards the damage, it could also be due to the other sources like electrical equipment and noise. Secondly, they cannot be used to estimate the crack severity. So, it is better to employ techniques like HT to identify and estimate the severity of the crack. Furthermore, by taking the FFT initially, helps in identification of the first modal frequency of the plate (28 Hz) and the frequency band in which to operate is distinguished.

The identification of the crack is carried out by applying the proposed methodology to the responses, as shown in Figure 6.7 for all the cases. The frequency change percent for the proposed and the direct method is found harmonic loading live transient signal. It is clear that the from proposed method is significant as compared to the direct method, thus enhancing the results.

6.3.3   Random Excitation

Three random tests were performed for analysis. The Figure 6.9 shows the random time waveform response of plate with 0.25 crack severity ratio.

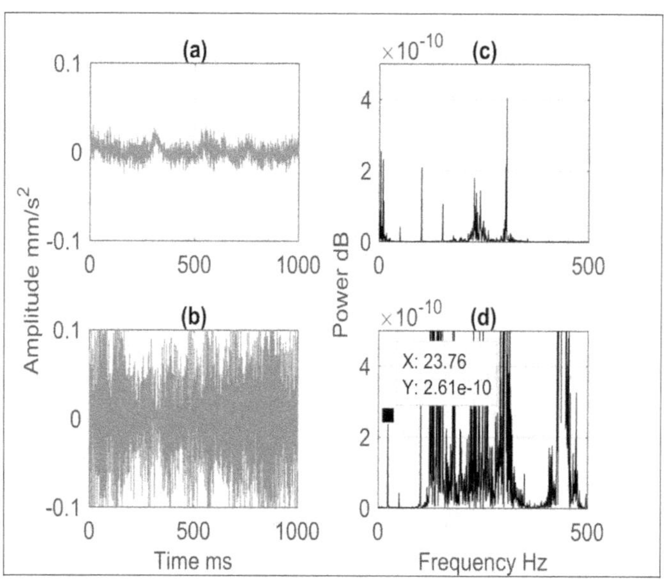

Figure 6.9: (a) Random input (b) Response (c) FFT of input (d) FFT of response

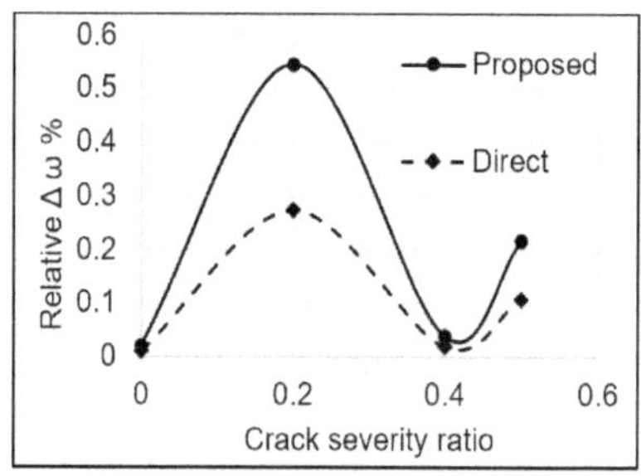

Figure 6.10: Relative frequency change under random excitation

The results of the random input as shown in Figure 6.10, successfully detects crack but severity is not in accordance with the harmonic or impulse excitation. This behavior may be due to the fact that experimentally, the opening and closing of the crack is not uniform. In case of higher severity like 0.50, the crack may remain open for a greater period as compared to the smaller crack, So, although the crack is detected but its increase in severity is not uniform.

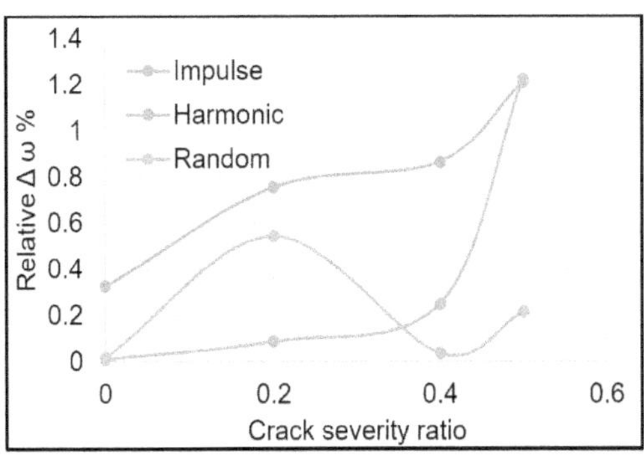

Figure 6.11: Comparison between different excitations on the experimental plate

The comparison between the three excitations on CFFF plate in Figure 6.11 shows that Impulse excitation provides good indication of damage whereas the severity is better estimated with Harmonic excitation as the trend is uniformly increasing till 0.4 severity ratio and then after 0.4 a sharp increase takes place showing that after 0.4, breathing phenomenon is increased in crack. For random excitation, although detection is possible but due to constantly varying amplitude and frequency, the trend is not regular so severity cannot be estimated at all.

6.4   VALIDATION OF STUDY

The validation of the study is carried out by comparing the graphs of analytical, simulated and experimental study under the different excitations. The results of the impulse, harmonic and random excitations have been shown in the Figure 6.12, 6.13 and 6.14, respectively.

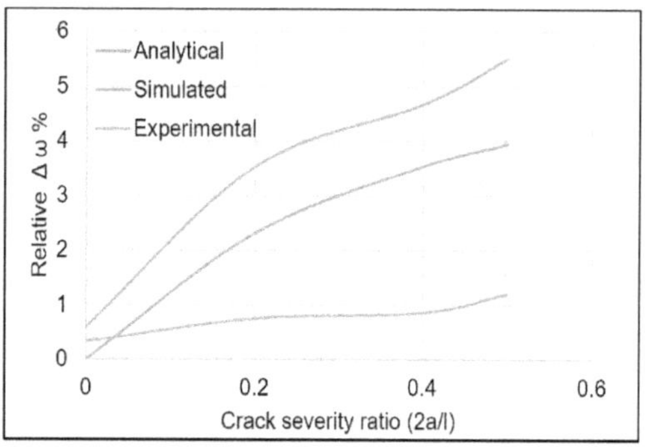

Figure 6.12: Comparison of analytical, simulated and experimental studies under Impulse excitation

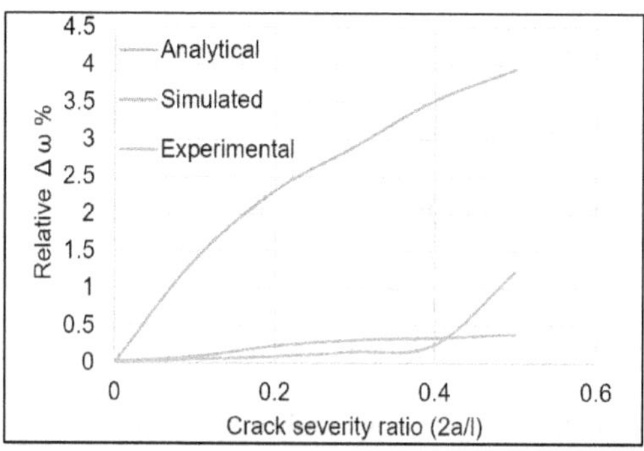

Figure 6.13: Comparison of analytical, simulated and experimental studies under Harmonic excitation

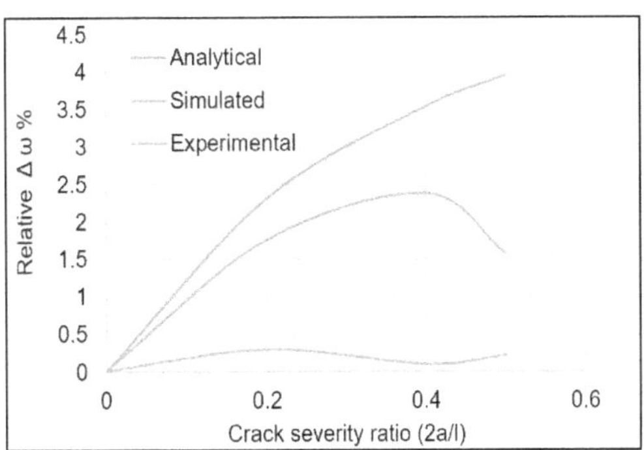

Figure 6.14: Comparison of analytical, simulated and experimental studies under Random excitation

Figure 6.12 shows that the analytical and simulated trend is almost same with some difference in amplitude only. This suggests that the simulated results have close agreement with analytical values in case of impulse excitation, whereas the harmonic and random excitations show greater deviation from the analytical values.

For harmonic case, Figure 6.13, the simulated and experimental values are almost the same. However, in experimental study, after 0.4 severity ratio a sharp increase in frequency change is observed, indicating higher nonlinearity. It can be said that for lab results harmonic excitation works best.

In case of random excitation, Figure 6.14, the simulated as well as experimental values shows deviation from the analytical results. Although the breathing crack is detected but its severity estimation is not reliable with random excitation. This may be due to the fact that the severity estimation is based on the relative difference of open and closed crack and for random both amplitude and frequency is constantly changing making it difficult to give a constant trend. Some further method must be applied to the random excitation to get good results.

6.5   RESULTS AND DISCUSSION

In this chapter, the experimental validation of the developed methodology for the case of CFFF plate has been investigated. As the CFFF BC was the worst case scenario from the analytical results, so the proposed method was implemented on this case.

The results are shown in the form of IFR in Figure 6.15. It has been observed that the performance of the proposed methodology is very simple and effective. The cracks are successfully detected for all excitations. The IFR shows the measure of severity. The lower its value, the more severe is the crack, hence showing its inverse relation. The IFR initially decreases slowly but as the severity ratio crosses 0.1, a sharp decrease is observed. This happens in impulse as well as harmonic loading. However, in random loading although the detection is valid, but the severity estimation follows a different trend. It may be possible since in random loading the wavelengths of the positive and negative halves of the waves vary, hence affecting the opening and closing behavior of the crack. So, the larger crack remains open most of the time thus showing lesser non-linearity.

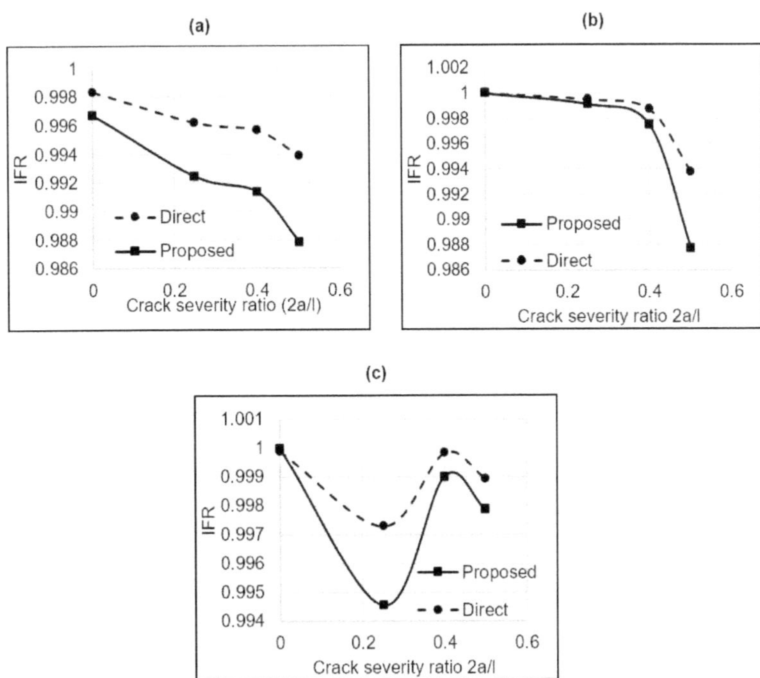

Figure 6.15: IFRs for CFFF plate under (a) Impulse (b) Harmonic (c) Random excitations

The dynamical behavior of the crack opening and closing is not uniform therefore the experimental values show different values of IFR (Figure 6.15) than that of the simulated model (Figure 5.31). However, when the graph between both are compared, it can be seen that the trend is the same. The difference of readings is not significant as the graph shows a relative trend among different crack severities.

7.CONCLUSION

7.1   CONCLUSION

The research work tells us about the crack identification methodologies that have been investigated by different researchers over the past few decades. Generally, open cracks in beams and plates have been researched and many techniques have been developed both analytically and numerically. The research has been done using vibrational study, sensors and signal processing techniques. Hence, showing that a sound knowledge of all three fields is required. The vibrational study of beams and plates gives us insight of the theories that have been explored in the past in the mechanical vibration field. The basic Euler Bernoulli theory for beams and the Kirchhoff-Love theory for thin plates has been quite popular. The existence of crack modifies the basic theories according to the presented problems. These developed theories help in crack identification. Although many mathematical models exist to identify open crack but study is still required for mathematical model of part-through breathing crack. In the last decade, many theories for breathing cracks have emerged in case of beams but for plates only a few are present. For the same reason, this research has been focused on part-through breathing crack with $2a$ length. Although when presenting a problem at hand, mathematical model plays the key role but for actual data from real life structure, data acquisition system is needed. The Data Acquisition System (DAC) requires sensors like accelerometers, input/output module and DAC software. The signal obtained from the system is then interpreted after filtering out noise and applying a specifically developed algorithm for identification. Many signal processing techniques exist for this but their application to find the solution for the particular problem is required. The aim of this thesis was to develop an analytical model of nonlinear cracks like breathing crack. The Kirchhoff-Love plate theory was used and the opening and closing of the part-through central crack was addressed. For this purpose, two frequencies, open crack frequency and closed crack frequency was extracted from the response signal (analytical and simulated) coming from the plate. The breathing crack frequency was somewhere between the two frequencies and calculated by equation 3.28. The responses from the cracked plate were then studied under different excitations so as to present the robustness of the method. Different crack severities and different boundary conditions were investigated and the trends developed. The comparison graph in Figure 5.29 highlighted the attributes of different excitations and drew important conclusions. HT proved to be vital tool in signal processing of the response. It tackled the nonlinearity of the signal by giving output in the form of both time and frequency. A bandpass filtered signal was used for applying HT and the IFs obtained were used for identification and severity estimation. The results from various scenarios showed that the proposed methodology was effective regardless of the type of excitation. Although the trend for various BCs was different for the severity estimation of the crack but identification was accurate. The increase in severity ratio resulted in an increase in relative frequency difference percentage. The graphs in Figure 5.20 showed that the proposed method significantly improved the results from the conventional method of crack identification. While the conventional method was unable to detect early crack, the proposed method detected crack at 0.05 crack severity ratio. The proposed method gives difference between frequencies of open and closed state of breathing crack whereas the direct method gives the difference of natural frequencies between the intact plate and cracked plate. The larger the difference, greater the severity of the crack hence showing direct relation between the two.

The crack identification is based on the difference between the frequencies of the open and closed state of the crack. Therefore, a minimum threshold value has been defined which checks the presence and absence of the breathing crack. As the intact plate frequency is considered to be that of closed crack state so it does not matter whether the intact frequency has changed over time or not. Hence, making the method baseline free. The study has not only carried out simulated experiments but experimental validation has also been conducted. The cantilever case has been explored for three different severity ratios and the trend developed for the frequency ratio, known as IFR in this study. It can be seen from Figure 6.15 that the ratio drops as the severity increases. All the different excitations show different trends. The impulse contributes to early crack detection while harmonic gives a linear slow increase till 0.4 severity and then sharp increase after 0.4. The prediction of crack severity would be easier from harmonic trend. The random excitation although detects breathing crack but prediction of crack severity is not feasible due irregular trend.

In this research mode I has been considered whereas operating machinery uses higher modes. So, this work can be applied on higher modes.

7.2   RECOMMENDATIONS FOR FUTURE

The study on breathing cracks in a plate has a huge scope as it is just an initial stage in identifying the crack. The localization of crack can be done in the future both mathematically and by simulations. The breathing crack at different angles and positions can also be investigated. Also, the work maybe extended to curved plates or circular plates.

The proposed technique can be tested with multiple breathing cracks located spatially at a different location and have contrasting open and close behavior (i.e. one crack opens while the other closes vice versa).

Plate on composite materials can be studied and smart materials can also be considered for crack identification. Moreover, the mathematical formulation has the capacity to be applied to higher modes. So, the work can be extended to higher modes for crack identification and severity estimation.

BIBLIOGRAPHY

[1] A. Motallebi, M. Fathalilou, and G. Rezazadeh, "Effect of the open crack on the pull-in instability of an electrostatically actuated micro-beam," *Acta Mechanica Solida Sinica*, vol. 25, no. 6, pp. 627–637, 2012.

[2] D. Goyal and B. Pabla, "The vibration monitoring methods and signal processing techniques for structural health monitoring: a review," *Archives of Computational Methods in Engineering*, vol. 23, no. 4, pp. 585–594, 2016.

[3] L. Chen, J. Xue, Z. Zhang, and W. Zhang, "Bifurcation study of thin plate with an all-over breathing crack," *Advances in Materials Science and Engineering*, vol. 2016, 2016.

[4] S. E. Khadem and M. Rezaee, "An analytical approach for obtaining the location and depth of an all-over part-through crack on externally in-plane loaded rectangular plate using vibration analysis," *Journal of sound and vibration*, vol. 230, no. 2, pp. 291–308, 2000.

[5] A. Israr, M. P. Cartmell, E. Manoach, I. Trendafilova, W. Ostachowicz, M. Krawczuk, and A. Żak, "Analytical modeling and vibration analysis of partially cracked rectangular plates with different boundary conditions and loading," *Journal of Applied Mechanics*, vol. 76, no. 1, 2009.

[6] H. Aftab, U. Baneen, and A. Israr, "Identification and severity estimation of a breathing crack in a plate via nonlinear dynamics," *Nonlinear Dynamics*, pp. 1–17, 2021.

[7] A. Rytter, "Vibrational based inspection of civil engineering structures," 1993.

[8] L. Pieczonka, A. Klepka, A. Martowicz, and W. J. Staszewski, "Nonlinear vibroacoustic wave modulations for structural damage detection: an overview," *Optical engineering*, vol. 55, no. 1, p. 011005, 2015.

[9] P. B. Nagy, "Fatigue damage assessment by nonlinear ultrasonic materials characterization," *Ultrasonics*, vol. 36, no. 1-5, pp. 375–381, 1998.

[10] P. P. Delsanto, *Universality of nonclassical nonlinearity*. Springer, 2006.

[11] K.-Y. Jhang, "Nonlinear ultrasonic techniques for nondestructive assessment of micro damage in material: a review," *International journal of precision engineering and manufacturing*, vol. 10, no. 1, pp. 123–135, 2009.

[12] N. P. Yelve, M. Mitra, and P. M. Mujumdar, "Spectral damage index for estimation of breathing crack depth in an aluminum plate using nonlinear lamb wave," *Structural Control and Health Monitoring*, vol. 21, no. 5, pp. 833–846, 2014.

[13] K. E. Van Den Abeele, P. A. Johnson, R. A. Guyer, and K. R. McCall, "On the quasi-analytic treatment of hysteretic nonlinear response in elastic wave propagation," *The Journal of the Acoustical Society of America*, vol. 101, no. 4, pp. 1885–1898, 1997.

[14] L. Ostrovsky and P. A. Johnson, "Dynamic nonlinear elasticity in geomaterials," *La Rivista del Nuovo Cimento*, vol. 24, no. 7, pp. 1–46, 2001.

[15] K. Worden, C. R. Farrar, J. Haywood, and M. Todd, "A review of nonlinear dynamics applications to structural health monitoring," *Structural Control and Health Monitoring: The Official Journal of the International Association for Structural Control and Monitoring and of the European Association for the Control of Structures*, vol. 15, no. 4, pp. 540–567, 2008.

[16] O. Giannini, P. Casini, and F. Vestroni, "Nonlinear harmonic identification of breathing cracks in beams," *Computers & Structures*, vol. 129, pp. 166–177, 2013.

[17] D. Broda, W. Staszewski, A. Martowicz, T. Uhl, and V. Silberschmidt, "Modelling of nonlinear crack–wave interactions for damage detection based on ultrasound—a review," *Journal of Sound and Vibration*, vol. 333, no. 4, pp. 1097–1118, 2014.

[18] A. Klepka, L. Pieczonka, W. J. Staszewski, and F. Aymerich, "Impact damage detection in laminated composites by non-linear vibro-acoustic wave modulations," *Composites Part B: Engineering*, vol. 65, pp. 99–108, 2014.

[19] U. Andreaus and P. Baragatti, "Experimental damage detection of cracked beams by using nonlinear characteristics of forced response," *Mechanical Systems and Signal Processing*, vol. 31, pp. 382–404, 2012.

[20] M. Dunn, A. Carcione, P. Blanloeuil, and M. Veidt, "Critical aspects of experimental damage detection methodologies using nonlinear vibro-ultrasonics," *Procedia engineering*, vol. 188, pp. 133–140, 2017.

[21] D. Joglekar, "Analysis of nonlinear frequency mixing in timoshenko beams with a breathing crack using wavelet spectral finite element method," *Journal of Sound and Vibration*, vol. 488, p. 115532, 2020.

[22] L. Guan, M. Zou, X. Wan, and Y. Li, "Nonlinear lamb wave micro-crack direction identification in plates with mixed-frequency technique," *Applied Sciences*, vol. 10, no. 6, p. 2135, 2020.

[23] J. Yin, Q. Wei, L. Zhu, and M. Han, "Nonlinear frequency mixing of lamb wave for detecting randomly distributed microcracks in thin plates," *Wave Motion*, vol. 99, p. 102663, 2020.

[24] E. Douka and L. Hadjileontiadis, "Time–frequency analysis of the free vibration response of a beam with a breathing crack," *Ndt & E International*, vol. 38, no. 1, pp. 3–10, 2005.

[25] T. B. Quy and J.-M. Kim, "Crack detection and localization in a fluid pipeline based on acoustic emission signals," *Mechanical Systems and Signal Processing*, vol. 150, p. 107254, 2021.

[26] N. Li, F. Wang, and G. Song, "New entropy-based vibro-acoustic modulation method for metal fatigue crack detection: An exploratory study," *Measurement*, vol. 150, p. 107075, 2020.

[27] B. Liu, J. Yang, and T. Gang, "Analysis of sound and vibration interaction on a crack and its use in high-frequency parameter selection for vibro-acoustic modulation testing," *Mechanical Systems and Signal Processing*, vol. 143, p. 106835, 2020.

[28] E. P. Carden and P. Fanning, "Vibration based condition monitoring: a review," *Structural health monitoring*, vol. 3, no. 4, pp. 355–377, 2004.

[29] A. Gandomi, M. Sahab, A. Rahaei, and M. S. Gorji, "Development in mode shape-based structural fault identification technique," *World Applied Sciences Journal*, vol. 5, no. 1, pp. 29–38, 2008.

[30] F. Barzegar, S. Mohebpour, and H. Alighanbari, "Multi-crack detection in general cross-section swept wings based on natural frequency changes," *Journal of Vibration and Control*, p. 1077546320964010, 2020.

[31] M. Elshamy, W. Crosby, and M. Elhadary, "Crack detection of cantilever beam by natural frequency tracking using experimental and finite element analysis," *Alexandria engineering journal*, vol. 57, no. 4, pp. 3755–3766, 2018.

[32] P. Cawley and R. D. Adams, "The location of defects in structures from measurements of natural frequencies," *The Journal of Strain Analysis for Engineering Design*, vol. 14, no. 2, pp. 49–57, 1979.

[33] O. Salawu, "Detection of structural damage through changes in frequency: a review," *Engineering structures*, vol. 19, no. 9, pp. 718–723, 1997.

[34] P. Gudmundson, "The dynamic behaviour of slender structures with cross-sectional cracks," *Journal of the Mechanics and Physics of Solids*, vol. 31, no. 4, pp. 329–345, 1983.

[35] M.-H. Shen and Y. Chu, "Vibrations of beams with a fatigue crack," *Computers & structures*, vol. 45, no. 1, pp. 79–93, 1992.

[36] W. Xu, Z. Su, M. Radzieński, M. Cao, and W. Ostachowicz, "Nonlinear pseudo-force in a breathing crack to generate harmonics," *Journal of Sound and Vibration*, vol. 492, p. 115734, 2021.

[37] R. Szilard, "Theories and applications of plate analysis: classical, numerical and engineering methods," *Appl. Mech. Rev.*, vol. 57, no. 6, pp. B32–B33, 2004.

[38] S. S. Rao, *Vibration of continuous systems*. John Wiley & Sons, 2019.

[39] T. Asakura, T. Ishizuka, T. Miyajima, M. Toyoda, and S. Sakamoto, "Finite-difference time-domain analysis of structure-borne sound using a plate model based on the kirchhoff-love plate theory," *Acoustical Science and Technology*, vol. 35, no. 3, pp. 127–138, 2014.

[40] S. P. Timoshenko and S. Woinowsky-Krieger, *Theory of plates and shells*. McGraw-hill, 1959.

[41] G. Warburton, "The vibration of rectangular plates," *Proceedings of the Institution of Mechanical Engineers*, vol. 168, no. 1, pp. 371–384, 1954.

[42] A. Ugural, *Stresses in plates and shells*. McGraw-Hill, 1999.

[43] A. W. Leissa, "The free vibration of rectangular plates," *Journal of sound and vibration*, vol. 31, no. 3, pp. 257–293, 1973.

[44] D. Dawe and O. Roufaeil, "Rayleigh-ritz vibration analysis of mindlin plates," *Journal of Sound and Vibration*, vol. 69, no. 3, pp. 345–359, 1980.

[45] S. Kitipornchai, Y. Xiang, C. Wang, and K. Liew, "Buckling of thick skew plates," *International Journal for Numerical Methods in Engineering*, vol. 36, no. 8, pp. 1299–1310, 1993.

[46] J. R. Rice and N. Levy, "The part-through surface crack in an elastic plate," 1972.

[47] R. Solecki, "Bending vibration of a rectangular plate with arbitrarily located rectilinear crack," *Engineering Fracture Mechanics*, vol. 22, no. 4, pp. 687–695, 1985.

[48] Y. S. Wen and Z. Jin, "On the equivalent relation of the line spring model: A suggested modification," *Engineering Fracture Mechanics*, vol. 26, no. 1, pp. 75–82, 1987.

[49] Z.-J. Zeng, S.-H. Dai, and Y.-M. Yang, "Analysis of surface cracks using the line-spring boundary element method and the virtual crack extension technique," *International journal of fracture*, vol. 60, no. 2, pp. 157–167, 1993.

[50] T. Bose and A. Mohanty, "Vibration analysis of a rectangular thin isotropic plate with a part-through surface crack of arbitrary orientation and position," *Journal of Sound and Vibration*, vol. 332, no. 26, pp. 7123–7141, 2013.

[51] K. Worden, *Nonlinearity in structural dynamics: detection, identification and modelling*. CRC Press, 2019.

[52] H. Long, Y. Liu, and K. Liu, "Nonlinear vibration analysis of a beam with a breathing crack," *Applied Sciences*, vol. 9, no. 18, p. 3874, 2019.

[53] Y. Ohara, T. Mihara, and K. Yamanaka, "Effect of adhesion force between crack planes on subharmonic and dc responses in nonlinear ultrasound," *Ultrasonics*, vol. 44, no. 2, pp. 194–199, 2006.

[54] V. Gusev, B. Castagnede, and A. Moussatov, "Hysteresis in response of nonlinear bistable interface to continuously varying acoustic loading," *Ultrasonics*, vol. 41, no. 8, pp. 643–654, 2003.

[55] A. Chatterjee, "Structural damage assessment in a cantilever beam with a breathing crack using higher order frequency response functions," *Journal of Sound and Vibration*, vol. 329, no. 16, pp. 3325–3334, 2010.

[56] T. Chondros, "The continuous crack flexibility model for crack identification," *Fatigue & Fracture of Engineering Materials & Structures*, vol. 24, no. 10, pp. 643–650, 2001.

[57] G. Yan, A. De Stefano, E. Matta, and R. Feng, "A novel approach to detecting breathing-fatigue cracks based on dynamic characteristics," *Journal of Sound and Vibration*, vol. 332, no. 2, pp. 407–422, 2013.

[58] G. Carr, M. Chapetti, *et al.*, "On the detection threshold for fatigue cracks in welded steel beams using vibration analysis," *International journal of fatigue*, vol. 33, no. 4, pp. 642–648, 2011.

[59] A. Bouboulas and N. Anifantis, "Three-dimensional finite element modeling of a vibrating beam with a breathing crack," *Archive of Applied Mechanics*, vol. 83, no. 2, pp. 207–223, 2013.

[60] D. Joglekar and M. Mitra, "Nonlinear analysis of flexural wave propagation through 1d waveguides with a breathing crack," *Journal of Sound and Vibration*, vol. 344, pp. 242–257, 2015.

[61] W. Zhang, H. Ma, J. Zeng, S. Wu, and B. Wen, "Vibration responses analysis of an elastic-support cantilever beam with crack and offset boundary," *Mechanical systems and signal processing*, vol. 95, pp. 205–218, 2017.

[62] K. Zhang and X. Yan, "Multi-cracks identification method for cantilever beam structure with variable cross-sections based on measured natural frequency changes," *Journal of Sound and Vibration*, vol. 387, pp. 53–65, 2017.

[63] J. Zhou, L. Xiao, W. Qu, and Y. Lu, "Nonlinear lamb wave based dort method for detection of fatigue cracks," *NDT & E International*, vol. 92, pp. 22–29, 2017.

[64] P. Liu, H. Sohn, S. Yang, and H. J. Lim, "Baseline-free fatigue crack detection based on spectral correlation and nonlinear wave modulation," *Smart Materials and Structures*, vol. 25, no. 12, p. 125034, 2016.

[65] T. G. Chondros, A. D. Dimarogonas, and J. Yao, "Vibration of a beam with a breathing crack," *Journal of Sound and vibration*, vol. 239, no. 1, pp. 57–67, 2001.

[66] M. I. Friswell and J. E. Penny, "Crack modeling for structural health monitoring," *Structural health monitoring*, vol. 1, no. 2, pp. 139–148, 2002.

[67] N. Pugno, C. Surace, and R. Ruotolo, "Evaluation of the non-linear dynamic response to harmonic excitation of a beam with several breathing cracks," *Journal of sound and vibration*, vol. 235, no. 5, pp. 749–762, 2000.

[68] M. Rezaee and R. Hassannejad, "Free vibration analysis of simply supported beam with breathing crack using perturbation method," *Acta Mechanica Solida Sinica*, vol. 23, no. 5, pp. 459–470, 2010.

[69] F. Semperlotti, K. Wang, and E. Smith, "Localization of a breathing crack using nonlinear subharmonic response signals," *Applied Physics Letters*, vol. 95, no. 25, p. 254101, 2009.

[70] K. Dziedziech, L. Pieczonka, P. Kijanka, and W. J. Staszewski, "Enhanced nonlinear crack-wave interactions for structural damage detection based on guided ultrasonic waves," *Structural control and health monitoring*, vol. 23, no. 8, pp. 1108–1120, 2016.

[71] G.-W. Kim, D. R. Johnson, F. Semperlotti, and K. Wang, "Localization of breathing cracks using combination tone nonlinear response," *Smart materials and structures*, vol. 20, no. 5, p. 055014, 2011.

[72] H. Ma, J. Zeng, Z. Lang, L. Zhang, Y. Guo, and B. Wen, "Analysis of the dynamic characteristics of a slant-cracked cantilever beam," *Mechanical Systems and Signal Processing*, vol. 75, pp. 261–279, 2016.

[73] Z. Parsons and W. J. Staszewski, "Nonlinear acoustics with low-profile piezoceramic excitation for crack detection in metallic structures," *Smart Materials and Structures*, vol. 15, no. 4, p. 1110, 2006.

[74] M. Ryles, F. Ngau, I. McDonald, and W. Staszewski, "Comparative study of nonlinear acoustic and lamb wave techniques for fatigue crack detection in metallic structures," *Fatigue & Fracture of Engineering Materials & Structures*, vol. 31, no. 8, pp. 674–683, 2008.

[75] C. Zhou, M. Hong, Z. Su, Q. Wang, and L. Cheng, "Evaluation of fatigue cracks using nonlinearities of acousto-ultrasonic waves acquired by an active sensor network," *Smart materials and structures*, vol. 22, no. 1, p. 015018, 2012.

[76] H. Hu, W. Staszewski, N. Hu, R. Jenal, and G. Qin, "Crack detection using nonlinear acoustics and piezoceramic transducers—instantaneous amplitude and frequency analysis," *Smart materials and structures*, vol. 19, no. 6, p. 065017, 2010.

[77] A. Rivola, "Crack detection by bispectral analysis," in *XIII Congresso Nazionale Associazione Italiana di Meccanica Teorica ed Applicata-AIMETA*, vol. 97, 1997.

[78] P. Cacciola, N. Impollonia, and G. Muscolino, "Crack detection and location in a damaged beam vibrating under white noise," *Computers & structures*, vol. 81, no. 18-19, pp. 1773–1782, 2003.

[79] V. K. Nguyen and O. Olatunbosun, "A proposed method for fatigue crack detection and monitoring using the breathing crack phenomenon and wavelet analysis," *Journal of Mechanics of Materials and Structures*, vol. 2, no. 3, pp. 399–420, 2007.

[80] K. V. Nguyen, "Comparison studies of open and breathing crack detections of a beam-like bridge subjected to a moving vehicle," *Engineering Structures*, vol. 51, pp. 306–314, 2013.

[81] A. Khatkhate, A. Ray, E. Keller, S. Gupta, and S. C. Chin, "Symbolic time-series analysis for anomaly detection in mechanical systems," *IEEE/ASME Transactions On Mechatronics*, vol. 11, no. 4, pp. 439–447, 2006.

[82] I. Trendafilova and E. Manoach, "Vibration-based damage detection in plates by using time series analysis," *Mechanical Systems and Signal Processing*, vol. 22, no. 5, pp. 1092–1106, 2008.

[83] D. Rezaei and F. Taheri, "Damage identification in beams using empirical mode decomposition," *Structural Health Monitoring*, vol. 10, no. 3, pp. 261–274, 2011.

[84] B. Wimarshana, N. Wu, and C. Wu, "Application of entropy in identification of breathing cracks in a beam structure: Simulations and experimental studies," *Structural Health Monitoring*, vol. 17, no. 3, pp. 549–564, 2018.

[85] X. Wang and N. Wu, "Crack identification at welding joint with a new smart coating sensor and entropy," *Mechanical Systems and Signal Processing*, vol. 124, pp. 65–82, 2019.

[86] Y.-h. Huang, J.-e. Chen, W.-m. Ge, X.-l. Bian, and W.-h. Hu, "Research on geometric features of phase diagram and crack identification of cantilever beam with breathing crack," *Results in Physics*, vol. 15, p. 102561, 2019.

[87] Z.-R. Lu, D. Yang, J. Liu, and L. Wang, "Nonlinear breathing crack identification from time-domain sensitivity analysis," *Applied Mathematical Modelling*, vol. 83, pp. 30–45, 2020.

[88] E. Z. Moore, K. D. Murphy, and J. M. Nichols, "Crack identification in a freely vibrating plate using bayesian parameter estimation," *Mechanical Systems and Signal Processing*, vol. 25, no. 6, pp. 2125–2134, 2011.

[89] J. He, X. Guan, T. Peng, Y. Liu, A. Saxena, J. Celaya, and K. Goebel, "A multi-feature integration method for fatigue crack detection and crack length estimation in riveted lap joints using lamb waves," *Smart Materials and Structures*, vol. 22, no. 10, p. 105007, 2013.

[90] M. Hong, Z. Su, Y. Lu, H. Sohn, and X. Qing, "Locating fatigue damage using temporal signal features of nonlinear lamb waves," *Mechanical Systems and Signal Processing*, vol. 60, pp. 182–197, 2015.

[91] J. Prawin, "Localization of breathing crack in beam-like structures using super-harmonic components based on singular value decomposition approach," *International Journal of Structural Stability and Dynamics*, vol. 20, no. 01, p. 2050014, 2020.

[92] J. Prawin and A. Rama Mohan Rao, "Reference-free breathing crack identification of beam-like structures using an enhanced spatial fourier power spectrum with exponential weighting functions," *International Journal of Structural Stability and Dynamics*, vol. 19, no. 02, p. 1950017, 2019.

[93] J. Prawin and A. R. M. Rao, "Nonlinear identification of mdof systems using volterra series approximation," *Mechanical Systems and Signal Processing*, vol. 84, pp. 58–77, 2017.

[94] J. Prawin and A. R. M. Rao, "A method for detecting damage-induced nonlinearity in structures using weighting function augmented curvature approach," *Structural Health Monitoring*, vol. 18, no. 4, pp. 1154–1167, 2019.

[95] M.-H. Tien and K. D'Souza, "A generalized bilinear amplitude and frequency approximation for piecewise-linear nonlinear systems with gaps or prestress," *Nonlinear Dynamics*, vol. 88, no. 4, pp. 2403–2416, 2017.

[96] J. Prawin and A. R. M. Rao, "Nonlinear system identification of breathing crack using empirical slow-flow model," in *Recent Advances in Structural Engineering, Volume 1*, pp. 1075–1085, Springer, 2019.

[97] K. Wang, Y. Li, Z. Su, R. Guan, Y. Lu, and S. Yuan, "Nonlinear aspects of "breathing" crack-disturbed plate waves: 3-d analytical modeling with experimental

validation," *International Journal of Mechanical Sciences*, vol. 159, pp. 140–150, 2019.

[98] Y. Shen and C. E. Cesnik, "Nonlinear scattering and mode conversion of lamb waves at breathing cracks: An efficient numerical approach," *Ultrasonics*, vol. 94, pp. 202–217, 2019.

[99] J. Xue, Y. Wang, and L. Chen, "Nonlinear vibration of cracked rectangular mindlin plate with in-plane preload," *Journal of Sound and Vibration*, vol. 481, p. 115437, 2020.

[100] M. Civera, L. Z. Fragonara, and C. Surace, "Nonlinear dynamics of cracked, cantilevered beam-like structures undergoing large deflections," in *2019 IEEE 5th International Workshop on Metrology for AeroSpace (MetroAeroSpace)*, pp. 193–202, IEEE, 2019.

[101] S. B. Yamgoué and T. C. Kofané, "Application of the krylov–bogoliubov–mitropolsky method to weakly damped strongly non-linear planar hamiltonian systems," *International Journal of Non-Linear Mechanics*, vol. 42, no. 10, pp. 1240–1247, 2007.

[102] Y. Hao, W. Zhang, and J. Yang, "Nonlinear oscillation of a cantilever fgm rectangular plate based on third-order plate theory and asymptotic perturbation method," *Composites Part B: Engineering*, vol. 42, no. 3, pp. 402–413, 2011.

[103] M. Najafizadeh, J. Mohammadi, and P. Khazaeinejad, "Vibration characteristics of functionally graded plates with non-ideal boundary conditions," *Mechanics of Advanced Materials and Structures*, vol. 19, no. 7, pp. 543–550, 2012.

[104] K. Mazanoglu and M. Sabuncu, "Flexural vibration of non-uniform beams having double-edge breathing cracks," *Journal of Sound and Vibration*, vol. 329, no. 20, pp. 4181–4191, 2010.

[105] A. Israr, *Vibration analysis of cracked aluminium plates.* PhD thesis, University of Glasgow, 2008.

[106] Y. Ma and G. Chen, "Natural vibration of a beam with a breathing oblique crack," *Shock and Vibration*, vol. 2017, 2017.

[107] A. Rivola and P. White, "Bispectral analysis of the bilinear oscillator with application to the detection of fatigue cracks," *Journal of Sound and Vibration*, vol. 216, no. 5, pp. 889–910, 1998.

[108] M. Feldman, *Hilbert transform applications in mechanical vibration.* John Wiley & Sons, 2011.

[109] F. W. King, *Hilbert transforms*, vol. 1. Cambridge University Press Cambridge, 2009.

[110] N. E. Huang, Z. Shen, S. R. Long, M. C. Wu, H. H. Shih, Q. Zheng, N.-C. Yen, C. C. Tung, and H. H. Liu, "The empirical mode decomposition and the hilbert spectrum for nonlinear and non-stationary time series analysis," *Proceedings of the Royal Society of London. Series A: mathematical, physical and engineering sciences*, vol. 454, no. 1971, pp. 903–995, 1998.

[111] M. Simon and G. Tomlinson, "Use of the hilbert transform in modal analysis of linear and non-linear structures," *Journal of Sound and Vibration*, vol. 96, no. 4, pp. 421–436, 1984.

[112] G. Tomlinson, "Developments in the use of the hilbert transform for detecting and quantifying non-linearity associated with frequency response functions," *Mechanical Systems and Signal Processing*, vol. 1, no. 2, pp. 151–171, 1987.

[113] A. Agneni and L. Balis-Crema, "Damping measurements from truncated signals via hilbert transform," *Mechanical Systems and Signal Processing*, vol. 3, no. 1, pp. 1–13, 1989.

[114] M. Feldman, "Non-linear system vibration analysis using hilbert transform–ii. forced vibration analysis method'forcevib'," *Mechanical Systems and Signal Processing*, vol. 8, no. 3, pp. 309–318, 1994.

[115] K. L. Lawrence, *ANSYS tutorial release 13*. SDC publications, 2011.

[116] J. Jingpin, M. Xiangji, H. Cunfu, and W. Bin, "Nonlinear lamb wave-mixing technique for micro-crack detection in plates," *Ndt & E International*, vol. 85, pp. 63–71, 2017.

[117] J. Zeng, H. Ma, W. Zhang, and B. Wen, "Dynamic characteristic analysis of cracked cantilever beams under different crack types," *Engineering Failure Analysis*, vol. 74, pp. 80–94, 2017.

[118] H. Luo, X. Fang, and B. Ertas, "Hilbert transform and its engineering applications," *AIAA journal*, vol. 47, no. 4, pp. 923–932, 2009.

[119] S. Loutridis, E. Douka, and L. Hadjileontiadis, "Forced vibration behaviour and crack detection of cracked beams using instantaneous frequency," *Ndt & E International*, vol. 38, no. 5, pp. 411–419, 2005.

[120] A. Raich and T. Liszkai, "Benefits of implicit redundant genetic algorithms for structural damage detection in noisy environments," in *Genetic and Evolutionary Computation Conference*, pp. 2418–2419, Springer, 2003.

[121] S. Hosseini-Hashemi, M. Fadaee, and S. R. Atashipour, "A new exact analytical approach for free vibration of reissner–mindlin functionally graded rectangular plates," *International Journal of Mechanical Sciences*, vol. 53, no. 1, pp. 11–22, 2011.

[122] S. S. Rao, *Vibration of continuous systems*, vol. 464. Wiley Online Library, 2007.

[123] J. L. Lagrange, *Théorie des fonctions analytiques: contenant les principes du calcul différentiel, dégagés de toute considération d'infiniment petits, d'évanouissans, de limites et de fluxions, et réduits à l'analyse algébrique des quantités finies.* Ve. Courcier, 1813.

[124] S. R. Bistafa, "On the development of the navier-stokes equation by navier," *Revista Brasileira de Ensino de Física*, vol. 40, no. 2, 2018.

[125] A. E. H. Love, "Xvi. the small free vibrations and deformation of a thin elastic shell," *Philosophical Transactions of the Royal Society of London.(A.)*, no. 179, pp. 491–546, 1888.

[126] A. Goldenveizer, J. Kaplunov, and E. Nolde, "On timoshenko-reissner type theories of plates and shells," *International Journal of Solids and Structures*, vol. 30, no. 5, pp. 675–694, 1993.

[127] S. S. Singh and S. Tomar, "Quasi-p-waves at a corrugated interface between two dissimilar monoclinic elastic half-spaces," *International journal of solids and structures*, vol. 44, no. 1, pp. 197–228, 2007.

[128] J. Prawin, K. Lakshmi, and A. R. M. Rao, "A novel singular spectrum analysis–based baseline-free approach for fatigue-breathing crack identification," *Journal of Intelligent Material Systems and Structures*, vol. 29, no. 10, pp. 2249–2266, 2018.

[129] J. Prawin, K. Lakshmi, and A. R. M. Rao, "A novel vibration based breathing crack localization technique using a single sensor measurement," *Mechanical Systems and Signal Processing*, vol. 122, pp. 117–138, 2019.

[130] J. Prawin and A. Rama Mohan Rao, "Vibration-based breathing crack identification using non-linear intermodulation components under noisy environment," *Structural Health Monitoring*, vol. 19, no. 1, pp. 86–104, 2020.

VITA

Hajira Aftab

E-mail: eng.hajira@gmail.com

Education:

Ph.D (continued) 2021

Department of Mechatronics and Control Engineering

University of Engineering and Technology, Lahore.

MS Mechatronics and Control Engineering 2012

University of Engineering and Technology, Lahore

BSc. Mechatronics and Control Engineering 2007

University of Engineering and Technology, Lahore

Percentage 82.183 %

Intermediate (Pre-Engineering) 2003

Defence Degree College Lahore.

Percentage 83. 36 %

Matriculation 2001

Ibn-e-Sina College Lahore.

Percentage 86. 35 %

Work Experience:

I have worked in Center of Vibration Testing facility (2020-2021) in Institute of Space Technology, Islamabad. Also, for my experimentation I attended training on Vibration Analysis and also on Failure Analysis (2019). I served as a Lab Engineer as well as Lecturer in Institute of Quality and Technology Management, University of the Punjab

(2013-2015).

Subject Interests:

Automation, Structural Health Monitoring, Signal Processing Techniques and Mechanical Vibration.